Would You
Baptize an
EXTRATERRESTRIAL?

ALSO BY GUY CONSOLMAGNO, SJ

The Heavens Proclaim: Astronomy and the Vatican

God's Mechanics: How Scientists and Engineers Make Sense of Religion

*Intelligent Life in the Universe? Catholic Belief and the
Search for Extraterrestrial Intelligent Life*

The Way to the Dwelling of Light: How Physics Illuminates Creation

Brother Astronomer: Adventures of a Vatican Scientist

*Turn Left at Orion: Hundreds of Night Sky Objects to See in a Home
Telescope—and How to Find Them* with Dan M. Davis

Worlds Apart: A Textbook in Planetary Sciences with Martha W. Schaefer

Would You Baptize an

EXTRATERRESTRIAL?

... and Other
Questions from the
Astronomers' In-box
at the
Vatican Observatory

BROTHER GUY CONSOLMAGNO, SJ
and **FATHER PAUL MUELLER, SJ**

IMAGE

NEW YORK

All rights reserved.

Published in the United States by Image, an imprint of the
Crown Publishing Group, a division of Random House LLC,
a Penguin Random House Company, New York,
www.crownpublishing.com

IMAGE is a registered trademark and the "I" colophon is a
trademark of Random House LLC.

Library of Congress Cataloging-in-Publication Data is available upon request.

ISBN 978-0-8041-3695-2
eBook ISBN 978-0-8041-3696-9

PRINTED IN THE UNITED STATES OF AMERICA

Jacket design by Michael Nagin
Jacket illustration: Burkhardt Studio, Inc.

10 9 8 7 6 5 4 3 2 1

First Edition

In memory of
Fr. Bill Stoeger, SJ
(1943–2014)

Contents

INTRODUCTION: The Dialogue Begins

GUY: Would you baptize an extraterrestrial?

That is one of the questions people ask us all the time here at the Vatican Observatory . . . along with questions about the Star of Bethlehem, the beginning and end of the universe, Galileo, Pluto, black holes, killer asteroids, and all the other topics astronomers always get asked about.

What is it about questions of this sort that raises such interest—and sometimes suspicion and fear—among so many people? Let's face it, most people know we're not likely to be running into any ETs anytime soon; nor is the exact nature of the Star of Bethlehem essential to any catechism or creed. But people care. They keep asking us. Why?

This book is about what's behind those questions.

PAUL: And this book is about what it's like when science encounters faith on friendly, mutually respectful terms.

Do you think we should reject any results of modern science that seem to disagree with the Bible? Do you think that the Bible has greater authority than science, and that biblical faith should always get the last word over science? If so, this might not be the book for you. (But read on!)

Do you think we should reject anything in the Bible that seems to be at odds with modern science? Do you think that science has greater authority than the Bible, and that science should always get the last word over biblical faith? If so, this might not be the book for you. (But read on!)

Do you think that *both* science *and* faith should be taken seriously, but you struggle with how to hold science and faith together, with integrity? Do you find yourself tending to keep science and faith isolated from each other, in separate, watertight compartments, but you wish that science and faith didn't have to "take turns" in your life? Then this book is for you. Read on!

GUY: The two of us writing this book are members of the research staff at the Vatican Observatory, the official astronomical research institute of the Catholic Church. I'm a scientist; I specialize in planetary physics and geology and especially the study of asteroids and meteorites. Paul's area of expertise is the history and philosophy of science—especially that of physics and astronomy.

We are both Jesuits, which means that we are members of the Society of Jesus, the largest religious order in the Catholic Church. I am a Jesuit brother, and Paul's a Jesuit priest. When someone asks Paul or me whether we'd baptize an extraterrestrial, we find ourselves stopping to think: should I answer from the perspective of a scientist, from the perspective of a philosopher, from the perspective of a Catholic priest or brother—or from the perspective of someone who is all those things at once and doesn't find any great conflict or contradiction in that?

PAUL: Our primary work is research. Most of our time is spent working in the lab, making observations, writing papers, and attending scientific conferences. So we're in regular contact with our colleagues at other research institutes and universities.

But since most of us at the Vatican Observatory are Jesuit priests or brothers, we are also in frequent contact with members of the general public—people who have questions and comments about science and faith, people who want to tell us about some discovery they've made or about some theory they've devised. And also people who just want to talk. We get a lot of e-mail!

GUY: I have more than seven hundred such e-mails in my files from the last five years. Some of the messages are a bit off-the-wall. But all of them come from people who want to take science and faith seriously, and many of our correspondents are people having trouble figuring out how to hold the two together with integrity and consistency.

This book is structured around a half dozen particular questions we've been asked time and again—questions that are interesting in themselves but that also tend to presuppose a conflict of some sort between science and religion.

We start with the question of Genesis versus the Big Bang, and we discuss how science and religion can have different but complementary ways of looking at the same issue. Then we discuss how scientific theories and ideas change and evolve over time—for example, we describe what happened recently when astronomers debated the status of Pluto as a planet. And we ask how religion can or should respond when science evolves and changes. To see how that might work—and what happened when it didn't work well—we examine the case of Galileo's encounter with the Church.

People of faith generally believe that God is somehow active in the world. We talk about how that can be, in a universe that seems to be regulated and governed by inexorable scientific laws. People of faith often believe that we are somehow important in

God's eyes. We discuss how that can be, in a universe that is vast and ancient beyond human imagining . . . and that seems doomed to come to an inglorious end, eventually.

And, yes, as the title of the book promises, we also talk about whether we would baptize an alien: what could the message of Christ mean in a universe of countless planets and, for all we know, countless other races of intelligent beings?

PAUL: When people ask us these sorts of questions, the presupposition often seems to be that there must be some sort of conflict between science and religion. Lots of people think that they're in conflict and that you have to choose one or the other.

GUY: The idea of an eternal war between science and religion has been a popular theme in books and journalism for a while. But it dates only from the late Victorian era, in fact.

PAUL: And it does not date from the time of Galileo, as so many people seem to think.

GUY: But two simple observations get lost in the noise. Science and religion have common historical roots—the war between them (if there is one) has not been eternal. And many people who do science are also religious. At least for them—as for the two of us—religion and science are not at war at all.

PAUL: So how do science and religion actually relate to each other? Do they have to operate in separate, watertight compartments? Is one of them supposed to complement and serve the other? Or is it possible to look at their similarities, differences, and connections not in terms of some preconceived notion of what *should* be the case, but by seeing how science and faith actually do—or don't—work together? The conversation we have in this book reflects how science and faith can talk to each other.

GUY: Which is one reason why we've written the book in the form of a dialogue—as a series of six conversations between the two of us, Paul and Guy. The six conversations are intended to re-create the sorts of conversations that we've often had with each other, with other Jesuits, and with people we've met through our work. To spice things up, we imagine each of the conversations as taking place in some particular physical setting. Some of the settings are real places, and some are purely fictional.

PAUL: Writing in the form of a dialogue is also a tip of the hat on our part to Galileo, who wrote some of his great books in dialogue form.

GUY: Of course, Galileo included his opponents among the characters and put words into their mouths. We don't do that.

PAUL: We just put words into each others' mouths!

GUY: Paul and I are very fortunate: We get to live and work with a group of Jesuit scientists who take both science and faith very seriously. We all work together in the lab, but we also pray together in the chapel. In our daily lives, we don't feel any particular conflict or tension between science and faith. Paul and I want to share with you our simple-but-complex daily experience of taking science and faith seriously while not keeping them separated in isolated compartments.

PAUL: I've studied pastoral theology, and Guy's had a bit of that, as well; that's part of our Jesuit training. But when it comes to relating science and faith, we don't claim to be able to teach you "How to" do it. Instead, we simply want to share with you the joy and hope—and fun—that we find in doing science and living faith. We hope that our hope and joy will be contagious!

GUY: So read on, and feel free to argue with us along the way. We'll certainly be arguing with each other!

Day 1: Biblical Genesis or Scientific Big Bang?

SETTING: THE ART INSTITUTE OF CHICAGO

GALLERY OF THE EARLY TWENTIETH CENTURY

PAUL: Today we're going to talk about the beginning of all things, Creation itself. So it's fitting that we're in Chicago, since each of us had a kind of personal beginning here in the Windy City. You did philosophy studies at Loyola University shortly after you became a Jesuit. And I earned my doctorate in the history and philosophy of science at the University of Chicago.

GUY: When I read the opening verses of Genesis, where "a wind from God swept over the face of the waters," I always picture myself standing by the shore of a stormy Lake Huron; that's where I'm from. But, ah, Lake Michigan and the Windy City . . . It's tough for this Detroiter to admit it, but I love Chicago. And this area, the Museum Campus down by the lakeshore, is like heaven to me. So many favorite places of mine: the Adler Planetarium; the Shedd Aquarium . . .

But if we're going to talk about the beginning of the universe, why are we at the Art Institute? Shouldn't we be at the planetarium, where we could watch a show that illustrates the Big Bang on its giant dome? Or maybe we should be at the Field Museum of Natural History, with its wonderful collection of dinosaur fossils and meteorites.

PAUL: Think of all those people who are always asking us, as Vatican astronomers, about science and religion and the beginning of the universe . . .

GUY: The ones who want us to choose between Genesis and the Big Bang?

PAUL: Yes. Most of them aren't scientists, so I don't think that addressing their questions in a scientific place—a planetarium, for instance—would be all that helpful.

So often we divide our lives up into separate camps, separate buildings, if you will: aquarium versus planetarium, work versus play, science versus religion, and so forth. Sometimes it's hard to move from one to the other. I want us to start out today in a place where science and religion can overlap. Here at the Art Institute, you can't help but see that there's more than one way to represent reality—more than one style of painting, you might say.

So humor me for a moment. Let's look at some of my favorite works of art.

GUY: Hmm . . . isn't that Grant Wood's *American Gothic*? The weather-beaten couple standing in front of a white clapboard house, the man holding a pitchfork staring directly out of the canvas, the woman giving him a dirty look . . .

PAUL: The painting may be all too familiar, but I like it. Though it's as realistic as a photograph, it still seems to tell you something about these two people that a photo wouldn't be able to capture.

Now, head down this hall to the European section, and compare *American Gothic* with this painting . . .

GUY: Picasso's *The Old Guitarist*. I guess that's an old man seated with a guitar, but he's depicted in an almost abstract way. The guitar is painted realistically enough, but the old man is shown in weird lines and angles and a funky blue color scheme. That's a whole different take on the human form.

PAUL: It looks more "modern" than *American Gothic,* though it was painted about thirty years earlier.

Both paintings depict old people with the tools of their trades. Both were painted at roughly the same time, in the early twentieth century. Both communicate something deep and true about humanity, intellectually and emotionally, in ways that a book or a homily couldn't.

But neither painting tries to show everything about its subject matter. Each painting selects and emphasizes only certain things and leaves out other stuff that'd be irrelevant or that'd get in the way. And you know what? Science does something very similar. Science involves selective observation—science involves paying special attention to certain things while ignoring others.

GUY: Like a painting I remember seeing once, of a couple of people sitting in a city diner: a guy with a hawk nose and a woman in red sitting next to him, as seen through the diner window from across the street late at night. You could never take a photograph like that; there'd be parked cars and telephone wires in the way. But every time I see that painting, I get hungry for some fried eggs and coffee.

PAUL: Edward Hopper's *Nighthawks*. That one is also here at the Art Institute, over in Special Exhibitions.

GUY: I always thought of you as a physicist-turned-philosopher. I didn't know you were such an art nerd.

PAUL: I'm no art expert, believe me. But get this: back when I was in college (too many years ago), I once spent ninety minutes standing in front of a painting at the Isabella Stewart Gardner Museum, in Boston, taking notes on a yellow legal pad. It was an assignment I had to do for a course in art history. The painting was Titian's *Rape of Europa*: Jupiter, in the form of a white bull, is carrying Princess Europa away across the sea, where he will have his way with her. As I stood there taking notes, other museum visitors started asking me questions about the painting, as though I were some kind of expert! At first I laughed it off. But after a while I started talking with people about the painting, telling them what I had noticed about it and what I thought of it. And they told me what they thought. It led to some spirited discussions and disagreements as to what the painting means, as to whether or not it is beautiful, and why.

So it's not only art experts who can recognize and talk meaningfully about what's beautiful. It's not only scientists who can recognize what's true. And it's not only ethicists who can recognize what's good. Of course, experts can help and guide us. But things that are really beautiful or true or good can be recognized and appreciated by regular folks—if we are willing to open our eyes and take some time.

GUY: So, Mr. Art Expert, where do we go next?

LATE NINETEENTH CENTURY: IMPRESSIONISM

PAUL: *A Sunday on La Grande Jatte—1884;* it's one of my favorites. Georges Seurat painted a scene with elegant French families in a Parisian park on an island in the Seine River, circa 1884. He used a technique called pointillism: instead of strokes of paint, he built up the image by adding colors dot by dot . . . anticipating digital-imaging techniques by over a hundred years.

GUY: I hate to admit this . . . but when I see this painting, the first thing I think of is *Ferris Bueller's Day Off*.

PAUL: Oh, yeah, the movie with Matthew Broderick! This painting is the centerpiece of a pivotal scene in the movie. A trio of high school kids play hooky, and they have various improbable adventures around Chicago on a nice spring day. When they come here to the Art Institute, one of them, an angst-ridden teenager named Cameron, stares for a long time at *A Sunday on La Grande Jatte—1884*. As he gazes more and more deeply into the painting, with a stricken look on his face, the Parisian scene disappears from view: it falls apart into a chaotic, random collection of colored dots. It's at that moment that Cameron realizes that his own life seems to be falling apart into meaningless bits and pieces.

I have a certain sympathy for Cameron; I had my own share of teenage angst! But what I see when I look at Seurat's painting isn't chaos. I don't see the world falling apart. What I see is the world being analyzed down to its smallest, most basic parts.

When I look at *A Sunday on La Grande Jatte—1884*, I keep flipping back and forth between seeing the whole scene, which is lots of people enjoying a beautiful day in the park, and seeing the little dots from which that scene is made up. But to me that doesn't mean the world is falling apart—it means that there is more than one way to see the world. One way is to see the big picture, the everyday world of common experience. Another way is to see the world as analyzed by science: a world of tiny atoms, of particles and waves, of fields of force—a world that can be described mathematically.

That's one way to get at relating science and faith: think of it as flipping back and forth between two different ways of seeing one and the same world. We can see the world through the eyes of science or through the eyes of faith.

When you see the world through the eyes of faith, you are often very much concerned with everyday experiences of what is right and good and beautiful. You are concerned with how your life hangs together and makes sense—or doesn't!

But when you see the world through the eyes of science, your concerns are different. You want to know how the world works and what it's made of, right down to its smallest pieces. The world as analyzed by science can seem disconnected from the world of everyday experience, just as the dots in Seurat's painting can seem disconnected from the larger image.

The trick is to get comfortable with the idea of flipping back and forth between two different ways of seeing. And the trick, also, is not to panic if one way of seeing omits something that the other includes, or emphasizes something that the other neglects.

GUY: So you can see this painting as a collection of dots, or you can see it as an image of people in a park. Both descriptions are true. If one of them is true, it doesn't make the other one false.

PAUL: Just like you, Guy, I had to study some philosophy, as part of my Jesuit training. It drove me crazy. I was a science guy, and I just couldn't figure out what it was that the philosophers were worried about. It wasn't that I couldn't understand their answers. I couldn't even understand their questions!

Finally, after a year and a half of frustration, I had a kind of breakthrough. One day I was puzzling over Aristotle and his theory that each thing should be understood as being a composite of "matter" and "form." Now, I thought I knew what "matter" was, from my study of physics: it's what stuff is made of. But I couldn't figure out why, in addition to matter, you would need something else called "form."

Then suddenly it hit me, like a ton of bricks . . . I was making an assumption: that knowing what something is means breaking it down to its smallest parts (atoms, quarks, whatever) and figuring out the physical laws governing how they interact. But Aristotle doesn't make that assumption. For him, breaking a thing down to its smallest parts is only one part of what it means to know it. For Aristotle, to really know something, you have to see not just its parts but also the whole.

GUY: So not just the dots, but also the picture that the dots make?

PAUL: Even more than that. For Aristotle, really understanding a thing means knowing what it's made of, what kind of thing it is, who made it, and why they made it—what's its value or meaning. For Aristotle, questions like that come up as part of doing science.

But in modern science, the questions are much narrower. Questions about the value or meaning or purpose of things don't come up in modern science—they are set aside. As a result, modern science has become much more focused and efficient: we can analyze and describe all sorts of things we see in nature in terms of objective, mathematical laws. But the fact that questions about value or meaning or purpose don't come up in modern science doesn't mean that those questions aren't important.

That's the kicker—that's what hit me, back when I was first studying Aristotle. As a human being, I may want to know who made a thing and why. I may want to know about the value and meaning of a thing, even if science considers such questions irrelevant.

Modern science is incredibly successful at doing what it does, from understanding how to control diseases to calculating what happens around a black hole. This can lead us to make the lazy assumption that the scientific way of seeing the world is the best way, or even the only legitimate way. That's the assumption I had fallen into while studying physics in college. But if you think that seeing the world through the eyes of science will show us everything that really matters, then you might as well think that you see everything that matters about *A Sunday on La Grande Jatte—1884* if you look at it dot-by-dot.

GUY: OK, I can see that. "Flipping back and forth" is a useful image to keep in mind whenever we get asked, "If you believe that God speaks to us through Scripture, then why don't you accept the story of Creation given to us in the Bible?" Maybe the people who ask us that aren't getting that there's more than one way to look at the picture. And so they end up wanting to treat the Bible like an astronomy textbook.

And the fact is, there's more than one way of looking at the picture even within the Bible itself. The Bible contains several different Creation stories. It just isn't possible for all of them to be *literally* true—they disagree with one another! So there has to be something else going on here. Since they can't all be true in the same way, that means you need to be able to develop some way to flip back and forth between different descriptions even within the Bible itself.

PAUL: But let's be careful here. We don't want to fall into the easy trap of saying, "Everything is relative; everything is true."

Earlier I was telling you about standing in front of a painting, arguing with passersby about what it meant and whether it was beautiful, and why or why not. It makes sense to have that kind of argument about the beauty of a painting, even a classic like *The Rape of Europa*. But it wouldn't make sense to have a similar argument about the truth of a classic law of science, like the law of pendulum motion.

GUY: Obviously, denying the truth of the law of the pendulum just makes you look silly. But are you saying that everyone, expert or not, has a right to accept or deny that a piece of art is beautiful?

PAUL: When it comes to things like pendulums, pulleys, and electric motors, the science is settled—not because some pointy-headed scientific expert tells us so, but because these things are tools that we rely upon and use every day and take for granted. But when it comes to questions of meaning and beauty, it's different. Those aren't questions that get "settled"; those are things that each generation has to appropriate for itself, building on what came before.

GUY: But wait, there are extreme realms where Newton's laws fail, where you need quantum physics or relativity. A pendulum swinging next to a black hole would act really strangely, compared to what you'd expect from Newton's laws.

PAUL: Yes, of course. When I say we *know* that portions of our science are settled, I mean that we know this within certain limits. In fact, we can use science to specify those limits. And within those limits, it's not simply that the science is merely probable, or that it might be true. Within those limits, it *is* true. And we know truths today that we didn't know five hundred years ago; our progress is cumulative.

Why is it that modern science is able to make cumulative progress? It's because the questions that science asks are so very narrow. That's the trade-off we make when we do science. We exclude important questions of beauty and meaning and purpose for the sake of being able to make some cumulative progress regarding our knowledge of the natural world.

Science proposes competing theories for what we see in the world—and our expectation is that, over time, one of those theories will win out and the others will be discarded. There will be cumulative progress. *But that's not what's going on with the different stories of Creation in the Bible.* They are not competing theories, where one is supposed to win out in the end. The Bible keeps all the various Creation stories, letting them stand side by side, even though they contradict one another.

GUY: The same thing happens in the Gospel stories. The Bible keeps four Gospels side by side, even though they give different and sometimes contradictory accounts of the life and death of Jesus.

It reminds me of when I was going through a Beatles craze and read every book I could find about them. Even when they told the same stories, those books all gave different takes on the group. And the ones written closest to the time when the Beatles were active weren't always the most accurate or the most insightful. But you need all those different points of view to get a good picture not only of who the Beatles were, but of who people thought they were.

PAUL: Your Beatles example is true, in that you need more than one vantage point to get the whole picture. But there's something else going on as well.

In science, the goal is to eliminate contradictions and inconsistencies over time: that's a priority. (That would be a priority in writing a true history of the Beatles, too.) But when it comes to faith, eliminating contradictions and inconsistencies isn't the main priority. The fact that the Bible keeps all those different stories right there side by side shows that there's something different going on in faith compared to science. The Bible is trying to do something else.

GUY: Maybe it's not really all that different. After all, in science, as you say, you throw away a theory that doesn't work. But it's never OK to throw away the data that doesn't fit the theory. (At least, not unless you have a really good reason to think that the data are flawed.)

Science books go out of date. We throw the old one away when a newer one comes out, when we have new theories. But we don't throw away our old data; we merely interpret them differently. New theories try to account for old data (and new data) in new ways.

Now notice, the Bible doesn't go out of date. Maybe it's the "data" of humankind's encounters with God; it's not the theory of how to interpret those events. Even after thousands of years, and several major changes in the way we understand the cosmology of the universe, the Bible continues to inform us. That's why you don't want to throw any of it away.

PAUL: No, I don't see it that way. Yes, the Bible includes data—stories of human encounters with the divine. But the Bible also includes layer upon layer of reflection upon that data. So the Bible is not just data: it is a particular account of data-plus-interpretation, which Christians take to be a privileged account.

And there continues to be new data and new interpretations. The encounter with God is always a personal encounter—God relates to each new generation in a new way, so more data and more interpretation are being generated all the time.

GUY: Bah! You might just as well say that a scientific experiment is a record of the scientist's "personal encounter" with her Geiger counter!

My point is this. In science, the raw data—the "history of the encounter," if you prefer—can often be contained in lists of numbers. But the story of a religious encounter can be subtle and hard to communicate. That's why the Bible resorts to so many different literary forms— history or poetry or storytelling. That's why it has to be interpreted. And those interpretations can themselves be contradictory, but each one can still help us see some important aspect of the truth . . . just as, in fact, in science we wind up learning and using different theories, or recording data in different ways to measure and explain the same phenomena.

Art depends on context in a way that science does not. To look at these paintings and get something out of them, you have to bring some knowledge with you to add to the painting. If you had never seen a guitarist, you wouldn't know what Picasso was painting. That's true, no matter how realistic the painting is—if you had never seen a pitchfork before, you'd have no idea what *American Gothic* was about, realistic as it is. And five hundred years from now, when the last diner has closed, whoever looks at that Hopper painting of the diner will never have the memories of rainy Friday evenings in the big city that I have.

PAUL: You're right: every human who practices a religion does so within some particular cultural context.

But context also matters in science. It's well documented that different scientists will tend to notice different kinds of data, or to find different kinds of data convincing, depending on what theory it is that they already hold, or depending on what sort of equipment they are more familiar with. If you expect to see light acting like a wave, chances are you'll more easily notice and be convinced by data that show light acting like a wave (rather than like a particle). If you are more familiar with cloud chambers than scintillation detectors, chances are, you'll tend to be more easily persuaded by data in the form of a clear picture of a single particle interaction, rather than by statistical data from many particle interactions. Scientific observation is theory-laden and practice-laden. Context matters in science. Science is done by people, with all their "personal" equations.

When I was in college, I worked part-time as a tech assistant for an experimental physics research group working at Bates Linear Accelerator, near Boston. I'll never forget my first visit to the Accelerator: my first encounter with "Big Science." One of my professors gave me a ride up to the Accelerator on a Saturday morning. The building didn't look all that impressive from the outside. But when I first came inside, I was completely wowed. Instruments and blinking lights everywhere! Huge magnets! Lead bricks galore! I was in hog heaven.

There was a group of serious-looking men (yes, all men back then) on the other side of the experimental hall, all of them in white lab coats, all gathered around looking at something. My professor walked over to them, and I followed. I was thinking: This is it, I'm encountering Big Science and people who do Big Science.

But when I came closer, I saw what it was they were doing. They were gathered around a black-and-white portable television ... watching a Road Runner and Wile E. Coyote cartoon.

GUY: What, they were analyzing the physics of the Coyote flying off the cliff?

PAUL: No, they were just taking a break, having a good time.

GUY: So much for Big Science.

PAUL: No, that's what got me hooked on science, all the more! That's when I realized: these are my people! Science is supposed to be fun! Later, when I was a high-school physics teacher, I would sometimes include Road Runner cartoons on tests and quizzes. I'd ask my students to identify by name the law of physics the Road Runner happened to be violating at a particular moment.

But here's a question to ask yourself ... if you tend to see faith and science as being in conflict, which of the two do you identify with the Road Runner? And which do you identify with the Coyote?

GUY: Oh, that's easy. Obviously, science is the Coyote. No question about it: trying to capture an elusive prey, using all the latest tools from the Acme of our technological age ... That's what we do in science. And it seems to work for a while, until you extrapolate a bit too far and find yourself in midair, with nothing to support you, and an anvil for a parachute.

PAUL: I agree, science is the Coyote. The Coyote takes science seriously, and he tries to put it to good use. Like the Coyote, scientists take observational data and physical laws seriously—in fact, they're pretty much the only things that science takes seriously. Like the Coyote, scientists are willing to suffer the consequences of that focus—they are willing to get refuted, and they take their lumps when the equipment doesn't work or when they don't get the theory quite right.

The Coyote has every right to be intensely frustrated by the Road Runner, who so often seems to get away with blithely disregarding the laws of physics. The Road Runner doesn't so much disobey the laws of physics as simply refuse to make them too much of a priority. It can be pretty exasperating for scientists when people of faith come barreling through, like the Road Runner, blithely disregarding what science holds to be important, and unembarrassed by the fact that their faith appears to have little or no supporting data. Why should the Road Runner (and faith) escape refutation, when the Coyote (and scientists) endure it all the time? Not fair!

GUY: But notice, it's only the "science" part of Genesis—its description of the Creation of the universe—that was flattened by later scientific understandings of how the world works. The question of *why* the world was made is still treated as well in Genesis as it is anywhere else.

Indeed, the history of cosmology, our attempt to come up with a picture of the universe and how it works and where it came from, involves one comic turn after another. The more we learn, the more we realize how ridiculous our previous attempts were. Yet every cosmology always starts out with perfectly logical deductions from observing the world around us.

After all, the "cosmology" of Genesis is merely an obvious extrapolation from what people could observe for themselves. Genesis, chapter 1, describes a flat Earth, with a dome overhead. That's just what you see when you walk outdoors. That's what the paintings in this museum show. Look, here, at this painting of a field at harvesttime.

PAUL: *Stacks of Wheat (End of Summer)*, by Claude Monet. Around 1890, it says here.

GUY: Forget the wheat; look at the horizon. When a painter like Monet paints a horizon, it is—well, horizontal. And notice the sky behind the stacks. That sky is not just overhead; it rises straight up from the horizon, like the side of a dome. In fact, every time I look up at the stars in a good dark sky (which is all too rare nowadays), even though I know in my head that all the stars are suns out there at a whole range of different distances from Earth, it still looks to me like they are spots of light on a solid dome arching over my head. That's not divine revelation; that's simple observation.

MCKINLOCK COURT: GREEK, ROMAN, AND BYZANTINE ART

GUY: Let's head downstairs to the ancient Greek section. That's the art I studied in college. I was thinking of being a classics and history major at one point, and I took one of those art-in-the-dark classes where we had to memorize a million slides of painted vases.

So take a look at the figures on one of these vases. *Kraters*, they're called, mixing bowls. That's where we get the word "crater" to describe the bowl-shaped holes on the surface of the moon. This one's from Athens, around 460 BC. You can see Zeus zapping a guy who ticked him off . . .

You can't tell from this painting if the characters know that they're on a round Earth, but at least they're on a round vase—if Zeus keeps walking, he'll just get back to where he's started. But, in fact, whoever painted this vase probably knew, as well as Monet knew, that the world was round. That was well understood by the fifth century BC in Greece.

But the stories in Genesis derive from well before that time. The Babylonians were the dominant culture back when Genesis was written. Their cosmology started with observation, too. They observed the stars and picked out patterns that looked like pictures to them: constellations. Many of them are the same constellations we use even today. They also saw, moving slowly through the stars of the constellations, seven brighter objects that they called "wanderers," or, in Greek, "planets."

Planets are up in the sky with the stars; you can see them yourself with your naked eye, just like the Babylonians could, even today with city lights . . . if you know where to look. The most obvious of those "wanderers" was the sun itself. Of course, the brightness of the sun overwhelms the sky, so you can't see what stars are around it. But as the seasons pass, you can see that different stars are visible at night, while other stars that used to be visible become lost in the sunlight. Eventually you can figure out which star-picture is being overwhelmed by the sun in any given month. There were twelve of these pictures in all that the sun (and the other planets) moved through; the sun itself moved through one such picture per month. Those stars are mostly in patterns that look like animals—Aries the Ram, Taurus the Bull, Scorpius the Scorpion, and so forth. This "zoo" became known as the zoo-diac, or "zodiac."

Of all the bright wanderers moving through the stars, the moon moves the fastest. It's in a different part of the sky from night to night, going through all twelve zodiac constellations in a single month. But again, after a while you can see a pattern in where it moves and how fast it moves. Its phases, crescent to half to full moon and then back to crescent, repeat every twenty-nine or thirty days, in such a regular way that you can count the days of the lunar cycle into a "moonth."

PAUL: I remember hearing about a little kid who was explaining, with great seriousness, that the Earth has seven or eight *different* moons. As he explained it, different ones show up on different nights: one is crescent shaped, one is half a circle, one is completely round, and so on. A pretty sharp observation on his part! After all, no one on Earth is able to observe directly that it really is the same moon from one night to the next. The moon goes out of our field of view for half the day.

GUY: Besides the sun and the moon, careful observation showed that there were five other bright "stars" that changed their places slowly, month by month, among the constellations. One was very elusive, seen only near the sun at sunrise or sunset, popping up for a few days and then quickly disappearing for a few months before coming back again. One was bright and beautiful, dominating the evening or morning sky. One was scary red, sometimes brilliantly so and other times more subdued. A golden one moved with majestic grace among the stars, spending a year in each zodiac constellation. And one, pale yellow, moved very slowly indeed, taking about two and a half years on average to move from one constellation to the next.

So there they were, the seven wanderers: the sun and the moon, plus starlike Mercury, Venus, Mars, Jupiter (our old friend Zeus, from the *Krater*), and Saturn, named respectively for the gods of speed, beauty, war, majesty, and old age, as befit the colors and motions of those planets. They had the names of gods because they looked and acted like those gods.

Those planets were all known since ancient times ... and somebody, long before recorded history, made a week of seven days, with each day named for a planet. Those names live on in many languages—see the names of the days of the week in Spanish, French, Italian—and even English still has Saturn-day, Sun-day, Moon-day. (The other days in English are named for Norse versions of those same gods.)

Now, it was also an observed fact of nature (observation again!) that things that move are alive. Thus, it was clear to those early observers that those planets—which move, as anyone could see—must also be imbued with some sort of life.

Indeed, even the pagan idea of spirits inside rivers and trees was based on such observation and deduction. Things move when they're alive. And you can see that big living things—animals—breathe. So moving things are livings things, and they have "breath"—which happened to be the same word they used for "wind" and for "spirit": in Latin, *anima,* the root of our word "animal." Anything that has an *anima* is an animal. No mysticism here; these were merely obvious observations of nature.

That was the best science of its day, back when the Babylonians were first inventing astronomy. That's the science that was presumed by the writer of Genesis, the best science of the day: the seven days of the week, the breath of the wind on an initial chaos of water, the separation of a flat world under a dome, with water above and below that dome, to explain why water rains down on you from out of the sky and then where the water goes once it disappears into the ground.

The new thing in Genesis was not the shape of the universe: everybody knew that; they could see that for themselves. What was new in Genesis was the idea that the universe was created deliberately, in an orderly fashion, by a God who existed distinct from that universe, already there even before its beginning. That's the theme that's common to all the different Creation stories you can find in the Bible, from the two different stories in the first two chapters of Genesis to the "creation without cause" described in the second book of Maccabees. (And that's what is markedly different from the Babylonian Creation myths, which taught that the world was a mistake, an accidental offshoot of other things the gods were doing.)

PAUL: Ah, that's helpful—I want to hang on to that idea. *What's new in Genesis is the idea of a universe that is created intentionally and rationally by a God who exists distinct from that universe and prior to it.* That's important.

Science is concerned with the causal laws of nature—with how things work in the universe, with what can and cannot happen in the universe. What you are saying is that the story in Genesis is not about a God who creates *within* the universe, but about a God who creates the universe from "outside." So whatever sort of "causality" God employs in creating the universe, it isn't the same sort of physical causality that we see operating around us in the world—not the kind of causality that science tries to describe and explain. In creating the universe, God is not using the laws of physics; God is *establishing* and *sustaining* those laws.

GUY: The Genesis flat-Earth cosmology worked fine if you never left Babylon, but the ancient Greeks who came later had done enough traveling to realize that this cosmology began to fall apart once you started traveling long distances by sea. For example, they'd seen how different stars became visible as you moved north to south; they were sharp enough to figure out that this meant they were moving around on a round ball of a world.

A round world also explained why the shadows cast by tall poles or obelisks showed different lengths, or no length at all, as you moved north or south. Around the year 240 BC, Eratosthenes worked out (correctly) the size of the Earth just from measuring how much the sun's shadow changed as you moved up the Nile between Syene (now Aswan) and Alexandria, Egypt. And, of course, the Greeks also recognized that they were seeing the round shape of the Earth during a lunar eclipse, when the sun and moon were on opposite sides of Earth: Earth cast a visibly round shadow in the sunlight falling on the moon. The size of the shadow, compared to the size of the moon, even gave them an idea of how big the moon was, and, from that, how far from Earth it lay.

A round world meant that the stars were not in a dome overhead, but fixed into a sphere enclosing the world. The planets were seen to move between us and the stars (sometimes you can see the moon passing in front of a star), so presumably each planet had its own sphere . . . which obviously was transparent, like crystal, to let the distant stars (and other planets) shine through. Perfectly reasonable, logical, and, indeed, rather cleverly worked out.

Planets did not move in a smooth, constant motion but sometimes doubled back on their paths for a while. These odd motions indicated that these spheres themselves had a complicated motion or—more clever yet—that the axis of each sphere was embedded in another invisible sphere that itself spun along a different axis with a different rate and orientation. But now the universe consists of a great many spheres, many of them completely invisible, all of them moving, and so all of them "animated" by some sort of *anima*—a spirit or soul.

Indeed, the motions of the planets were as complicated as the movements of human history. Maybe there was a connection? Maybe, if you could predict the motions of the planets, you could also predict what was going to happen next week in the stock market, or who was going to be the next emperor. That'd be useful to know. So the pundits of those days put a lot of effort into working out how to use mathematics to predict the positions of the planets.

C. S. Lewis outlines all this in wonderful detail, in his book *The Discarded Image*. As he describes it, the way they pictured it, "Each sphere, or something resident in each sphere, is a conscious and intellectual being, moved by 'intellectual love' of God." Lewis notes that "the planetary Intelligences, however, make a very small part of the angelic population which inhabits the vast aetherial region between the moon and the Primum Mobile."

PAUL: So they thought each planet was alive?

GUY: Not exactly. Each planet was a bright spot embedded in a transparent sphere, but the motions of those spheres (which we see in the motions of the planets) were caused by spiritual beings, whom you could not see directly: "planetary Intelligences" or angels.

These beings were classified by names such as Thrones and Dominions. The last and lowest of the planets is the moon; once you get to the regions below the moon, you find yourself in the realm of "the aerial beings, the daemons." In some sources, the daemons could be good or bad; other cosmologists of those times divided the good daemons into the upper air, the bad ones into the lower air. By the time of the Middle Ages, Lewis tells us, "the view gained ground that all daemons alike were bad; were in fact fallen angels, or 'demons.'"

And, in fact, this is only the beginning of the census of all the different kinds of inhabitants of the universe as understood in the medieval cosmology, a complexity that is only faintly echoed in modern fantasies such as *The Lord of the Rings*.

This cosmology operated, with modifications, from the time of the ancient Greeks up until the seventeenth century and the scientific revolution. It's how everyone assumed the universe worked, during the time of Christ. A few hundred years after Christ, however, we come to Ptolemy, a Greek living in Egypt under the Romans. Using the records of planetary positions recorded for hundreds of years by the Greeks and Babylonians, he devised a geometrical system of circles, and circles within circles, that was able to reproduce where the planets had been and to predict where they were to be found at any given time in the future. And it worked! Sure, he assumed the Earth was at the center; that's the simplest thing to assume. And if you feel contemptuous of anyone who followed that system, I challenge you to try to come up with a system as good as his. Without analytical geometry. Without algebra. Without Arabic numerals.

PAUL: So on one hand, there was the description of the universe as a set of spheres within spheres, which is generally credited to a Greek from the fifth century BC named Eudoxus. That was completely different from the flat-Earth, domed picture of the universe found in Genesis. And both of them were different from the nested epicycles of the Ptolemaic system. But people were able to live with all these different cosmologies at the same time. This was in part because Ptolemy's system wasn't taken to be a description of the real structure of the heavens; it was taken to be a mere mathematical device for calculating the planets' positions.

GUY: But it's clear from the writings of the Church Fathers that some early Christians did worry about how to make the various world-pictures fit together. Here's what Saint Augustine had to say about the matter, in his commentary on the book of Genesis, dating from AD 400. His opening paragraph concludes that "No Christian will dare say that the narrative [of Genesis] must not be taken in a figurative sense. For Saint Paul says: 'Now all these things that happened to them were symbolic.'" (I'm quoting here from John Hammond Taylor's translation of Augustine citing Saint Paul. Nowadays we interpret this passage from Saint Paul differently; but the point is not what Saint Paul actually said, but what Augustine understood him to be saying.)

And after nineteen chapters of interpreting Genesis figuratively, in the light of then-modern physics, Augustine concludes by writing (again quoting from Taylor's translation):

> Usually, even a non-Christian knows something about the Earth, the heavens, and the other elements of this world, about the motion and orbit of the stars and even their size and relative positions, about the predictable eclipses of the sun and moon, the cycles of the years and the seasons, about the kinds of animals, shrubs, stones, and so forth, and this knowledge he holds to as being certain from reason and experience. Now, it is a disgraceful and dangerous thing for an infidel to hear a Christian, presumably giving the meaning of Holy Scripture, talking nonsense on these topics; and we should take all means to prevent such an embarrassing situation, in which people show up vast ignorance in a Christian and laugh it to scorn.

PAUL: Augustine's point is this: Since humans are capable of gaining true knowledge about the world by doing science, science must be taken into account when interpreting Scripture. If there's a passage of Scripture that seems to contradict what we know from science, then that passage should not be interpreted literally—it should be interpreted figuratively.

Of course, what Augustine took to be "knowledge held certain from reason and experience" was a cosmology that we now know is completely wrong. Science advances.

GUY: Saint Paul is even more direct. In several places in his letters, such as to the Ephesians or to the Colossians, he mentions the role of Jesus compared to the "thrones, dominions, powers" of the universe. Modern readers sometimes think that Paul is talking about the political rulers of his day. But in fact those terms are the descriptions of the spirits that were supposed to be moving the spheres of the universe. In essence, Paul is talking about the pagan gods. And what does he say about them? That Jesus overrules all of them. In the most dramatic instance of this, Colossians 2:15, he describes that Jesus "has stripped the sovereignties and the ruling forces, and paraded them in public behind him in his triumphal procession!" Like a conquering general, says Saint Paul, Christ has destroyed the power of the planets to rule our lives. We don't have to be afraid of them anymore.

And that's important, because you have to remember the power this old cosmology had over the human psyche. It was not a bunch of merely *invisible* or "spiritual" forces ruling the planets and presumably our lives; it was based on stuff that anyone could see for themselves.

You don't believe in the gods? I can take you outside at night and show them to you in the sky! You don't believe in their power? Then how else do you explain the ebb and flow of the tides, or the ebb and flow of the seasons? You don't think that we're sitting at the bottom of the chain of Creation, with perfect, unchanging motions visible above us? Then why is it that down here on Earth stuff inevitably runs down and comes to a stop? The pyramids can be seen to be rotting away in the desert, but the planets continue in their orbits, forever unchanging. And you don't believe in hell? Let me take you to the island of Vulcano, off the coast of Sicily, where you can smell the fire and brimstone coming out of a huge hole in the ground.

This stuff was real. Anyone could see it. By comparison, modern science is full of weird ideas that are much farther from everyday experience—for example, unseen "natural laws" that control the universe and the mystical, inexplicable force (called "gravity") that somehow holds the stars together and guides the planets in their courses.

Though Saint Paul tries to convince his readers that the old cosmology should have no power over them, people would still cling to it for another sixteen hundred years. It's all mixed together in the popular imagination with the far more rational metaphysics of Aristotle, and with the Christianized understanding of his philosophy produced by Saint Thomas Aquinas. It underlies all of the art, literature, and popular music of the times. You can't read Dante or Chaucer without recognizing it as the bones on which their stories are built.

And you can understand why people were reluctant to throw it over during the time of Copernicus and Galileo. The medieval account seemed to fit with regular daily experience. Copernicus and Galileo were asking people to give up their taken-for-granted picture of the structure and shape of the universe and accept a new picture in which the Earth is moving. But it sure doesn't feel as though the Earth is moving. And no physical experiment performed in Galileo's day sufficed to prove that the Earth really was moving.

You had to have some pretty strong proof to get people to toss out what seemed obviously true and start over again. When you read the debates in the literature during the seventeenth century, as people were coming to grips with this new physics, you wind up with great respect for the serious arguments put forth by both sides. It really wasn't until Newton's laws that the new cosmology finally made coherent, systematic sense. And the final edition of his *Principia* wasn't published until a hundred years after Galileo and nearly two hundred years after Copernicus.

EUROPEAN MEDIEVAL ART

PAUL: One of the interesting things to notice about the medieval world-picture is how it shows up in paintings from that era. Let's head back upstairs, to the medieval section of the Art Institute, so I can show you what I mean.

Here's a lovely diptych from 1275, two wooden panels with images intended for a church altar. On the left is an image of the *Virgin and Child Enthroned,* and on the right is a depiction of *The Crucifixion.* Both are beautiful representations of Christ on Earth. But from the style it is clear that these are not intended to be seen as snapshots of particular events that happened at a particular place and time. Each image is a timeless representation. His mother at left and his disciples on the right are depicted alongside Christ because they were important people, not because they just happened to be photobombing the scene. What is being represented is an enduring truth about their relationship to Christ, not a particular historical event.

GUY: And since different cities might have one or another of those saints as their patron, I suppose a painting like this might also have a political message—whose saint stands closest to The Boss?

PAUL: But now, let's walk around to the other side of this floor of the museum, moving ahead a few centuries. With the Renaissance, paintings start to represent particular moments in time, using the tricks of perspective to create a sense of space and to place the actors in realistic locations. Over here, David Teniers's painting of *Abraham's Sacrifice of Isaac* from 1664 captures Abraham with his sword above his son's head, about to strike and sacrifice him, at the very instant when the angel intervenes. You can see the angel's hand on the sword, stopping it in its downswing.

In that same era, Newton and Leibniz were coming up with

the mathematics of calculus, which allows you to treat reality dynamically. Time and space become dynamic variables that can be differentiated into points and instants, and ongoing changes can be integrated over time.

GUY: What's more, it was Newton, tracing the fall of an apple and the orbit of the moon with the same laws of gravity, who finally overcame the assumption that the physical laws governing the Earth below were different from the laws controlling the heavens above. For Newton, everything in the sky and everything on the ground followed the same rules of motion.

But even Newton invoked the existence and action of God to fill a "gap" in physical theory. Newton realized that the interaction of one planet with another might well perturb the stable simplicity of any given planet's orbit. He wondered how planetary orbits could remain stable over time, despite these regular perturbations among themselves. This was a gap in explanation that needed to be filled. Newton saw no possible way to explain the stability of orbits other than to invoke the intervention of God. He supposed that, like the hand of the angel in that painting, the hand of God intervened at intervals to correct or fix the orbits of the planets, to keep them stable.

PAUL: For Newton, the fact that God had to intervene to keep the planetary orbits stable—the fact that God had to fill a "gap" in physical theory—counted as a proof of God's existence. Apart from that lone exception, Newton saw the world as operating according to physical laws that were exact, inexorable, and exceptionless. From Newton's perspective, only an all-powerful and all-knowing mathematician-God could bring about and maintain such a well-ordered universe. So it wasn't as though there was a split between science and religion. Rather, science was seen as underwriting religion.

GUY: To quote Alexander Pope, "God said 'Let Newton Be' and all was light."

PAUL: But Newton's way of proceeding turned out to be theologically treacherous. If you think that God's job is to fill the "gaps" in physical theory, to explain things that science can't explain, then over time, as science advances, God gets put out of His job. As the "gaps" get filled by advances in science, the "proof" of God's existence gets turned on its head: instead of there being gaps that show the need for God, the disappearance of the gaps implies that there's no need for God, after all. The Jesuit theologian Michael Buckley, in his book *At the Origins of Modern Atheism*, has argued that precisely this pattern of thought destroyed many people's faith in God.

GUY: The French mathematician Pierre-Simon Laplace, armed with a hundred years' worth of development in mathematics beyond what Newton knew, developed a description of planetary orbits far more elaborate than Newton's, which was able to account naturally for the stability of planetary orbits over time. The story goes that when Laplace explained his theory to Napoleon, the emperor interrupted him to ask what role God had in maintaining the stability of the orbits. Laplace is said to have replied, "I have no need for that hypothesis."

PAUL: By invoking God's intervention to fill a "gap" in physical theory, Newton was effectively making God into a kind of scientific hypothesis—he was treating the question of the existence of God as something that could be disproved scientifically. What Laplace said to Napoleon was correct: there is no need for the scientific hypothesis that God intervenes to maintain the planets in stable orbits. The mistake was to treat the question of the existence of God as if it were a scientific hypothesis, subject to empirical testing.

If you are treating the question of the existence of God as if it were a scientific hypothesis, you're making God into just one more force or phenomenon in the universe, alongside all the others. But that makes God into a mere nature-deity, responsible if not for thunder and the growth of crops, then at least for the primordial Big Bang and the growth of the universe. That's not the Christian conception of God. God is not a phenomenon or cause operating *within* nature. The Christian God is *super*natural, outside space and time. God underwrites the existence and order of nature, from outside.

GUY: There is an irony in Laplace's dismissal of Newton's concern, by the way. In the two hundred years since his Napoleonic quip, mathematicians have discovered chaos theory, and astronomers now understand that planetary orbits can, indeed, be chaotic. There is good evidence that the early orbits of planets were unstable. Just because the solar system has been relatively quiescent for the past four billion years doesn't mean it was always that way.

Immanuel Kant carried the Newtonian view of the universe to new lengths in his book *Universal Natural History and Theory of the Heavens* in 1755. He argued that certain nebulous clouds of light seen by telescopes were, in fact, galaxies, or "island universes" just like our own Milky Way. Though his suggestion was hotly debated well into the early twentieth century, it served to expand the size of the universe to previously unimagined distances across the cosmos.

This whole picture seemed to hang together . . . until the end of the nineteenth century, when Maxwell's new mathematical description of light, and a whole series of experiments trying to understand how light carried energy and propagated itself as a wave, eventually showed that Newton's physics was no longer sufficient. It worked well for everyday affairs, but problems came into play in its treatment of light and radiation. Its predictions were logically inconsistent, and they didn't match the results of the experiments. That triggered the birth of relativity and quantum physics, the pillars of modern physics.

PAUL: The collapse of classical physics at that time continues to shake our understanding of how reality works. We really haven't completely come to grips with it yet. For one thing, quantum theory and general relativity aren't consistent with each other. Though both of them work astonishingly well, strictly speaking they can't both be true. You might say that we're in the position of the Coyote, who has obtained two of Acme's best contraptions, which, if used in tandem, would allow him to catch the Road Runner. But Coyote finds that, though each contraption works astonishingly well when used alone, the two contraptions just can't be made to work together. That's general relativity and quantum theory: each one works great alone, but it's hard to make them work together.

GUY: Einstein's General Theory of Relativity, first published in 1916, was the first big step. In this theory, Einstein proposed that space, time, matter, and energy are all interrelated. He suggested that the force of gravity could be understood as how mass warps the interrelated dimensions of space and time. A key point of his theory was dramatically confirmed in 1919 when Arthur Eddington observed a star very near the sun—he looked during an eclipse, so the sun was blocked off and the star was visible—and he saw that the apparent position of that star was slightly shifted from where it should have been, according to the maps made when the sun wasn't nearby to bend its rays. When the rays of light from the star passed close by the sun, they had been bent by the sun's gravity; which is to say, the shape of space near the sun had been warped by the sun's mass. Einstein's prediction was different from what Newton's laws had predicted, and Einstein, not Newton, was shown to be correct.

Given that such a warping should tend to attract all matter together, this observation revealed a paradox. Ever since Aristotle, at least, it had been assumed that the universe was essentially unchanging and eternal. With Copernicus and Newton, modern physics made the even broader assumption that the universe is essentially the same everywhere and always. We assume that the same laws of physics, and more or less the same physical conditions, can be found in the most far-off distant galaxy as you might expect to find here in our own solar system, operating in the same way in the past and in the future. So, given all that infinite time, why hadn't all of this mass in the whole universe had time to pull itself together into one point?

Einstein suggested there could be another force that existed, one previously unknown, that held mass apart and allowed the universe to be divided into individual galaxies, stars, and planets. He noted that such a force could be introduced into his equations as a "cosmological constant."

But then in 1922 the Russian physicist Alexander Friedmann showed that Einstein's equations were also compatible with a universe that was expanding, and he demonstrated mathematically how that expansion could be related to the possible curvatures of space. Friedmann proposed that if the universe started out at the beginning with a sufficiently large expanding velocity, it could continue to expand indefinitely, even against the force of gravity, thus removing the need for a cosmological constant.

The concept of building an entire cosmology based on such an expansion of the universe from a single, highly dense quantum state is generally attributed to a 1927 paper by the Belgian mathematician and astrophysicist Georges Lemaître, who suggested that the universe grew from one immensely dense energetic point he called the "cosmic egg," and that you could even calculate the point in time when this cosmic egg was hatched: the beginning of the universe.

Today we call it the Big Bang.

PAUL: This is a point that always confuses people. They're always asking, "What is it that the Big Bang is expanding into?"

GUY: You can't think of this as material spreading out into an empty void. Instead, space itself is expanding. All of space and time, the entirety of the universe, was contained within the initial state, and so there is no physical meaning in speaking of anything "outside" this state. Nor can you necessarily speak of a physical time "before" this expansion.

It is also important to realize that the "space" that is expanding is not the space between you and me, or between our sun and other nearby stars, or even between our galaxy and the nearby Androm-eda galaxy. Our local gravity is strong enough to hold things to-gether against this tendency to expand. But that's not the case with the space between clusters of galaxies.

PAUL: So we have planets, like Earth and Mars, orbiting a star like our sun. Billions of stars like our sun, all orbiting around a com-mon center, make up a galaxy. The Milky Way is what we call the galaxy that we're inside; the Magellanic Clouds and the Androm-eda galaxy are other nearby galaxies, which you can actually see with your unaided eye if you know exactly where to look and the night is dark.

GUY: Right. But galaxies are not spread uniformly throughout the universe; they come in clumps, called clusters. Note that all the galaxies you mentioned are part of what modern astronomers call, unimaginatively, "The Local Group." However, even with just a good amateur's telescope you can begin to see other groups of gal-axies in the nighttime sky . . . there's a cluster of galaxies in the con-stellation of Ursa Major, and another cluster in Virgo, and another cluster in Leo . . .

Now, even the galaxies within a cluster are bound to one an-other by gravity and use their gravity to hold their space together. But what we observe is that various clusters of galaxies are, indeed, separated by much greater distances from one another. In the spaces between them, gravity is weak enough that the expansion of space can occur there. It's the relative motions of galaxy clus-ters that Edwin Hubble first observed in 1929, and which we can determine with far higher accuracy now, when we look to measure the expansion of the universe.

PAUL: This theory of an expanding universe was proposed at a particularly interesting time in the history of physics. Along with Einstein's relativity work and Hubble's observations of the expanding universe, the 1920s also saw the development of quantum physics, which describes the physics of atoms and subatomic particles. In the quantum realm, the classical physical picture of a mechanical universe, made up of objects with well-defined locations and trajectories, just doesn't work: the electron in a hydrogen atom has no well-defined trajectory, so it's impossible to specify exactly where it is or how fast it's moving at a given moment.

The confluence of these three breakthroughs—relativity, quantum physics, and the expanding-universe observations—motivated new cosmological theories and gave scientists the tools they needed to predict the kinds of effects that should be observed if the theories were correct. But the strange physics needed to describe nature at this level raises a big problem we're still trying to come to grips with, regarding modern science: it's often just too counterintuitive for regular folks to grasp.

CONTEMPORARY ART, 1945–1960

PAUL: To get at what I mean, let's go over to the Modern Art galleries. This is art that was starting to be made at just the same time Einstein and Bohr and all the others were coming up with modern physics. And this is one of the main reasons I wanted to come to the Art Institute today.

So, here's what happens to me when I look at the artwork around us. I'm fascinated by works of art that show me things I can recognize. That engages me. But I just don't "get" abstract art. It leaves me cold.

Remember the paintings I showed you at the beginning of our visit? Think of Picasso's *The Old Guitarist*. OK, it's funky, but I like that. Picasso's way of representing that old man pushes me well beyond my normal experience—I've never seen a guitar player who is blue, or one with such an angular form—but at least I can recognize that it's an old man playing a guitar.

Picasso's painting represents something familiar—an old man playing a guitar—in a new way: blue and angular. Some art historians say that what Picasso did was to play with dimensions, to force his subjects into the canvas, so that the flatness of the canvas is the reality and we recognize that our idea of perspective is the illusion.

Galileo and Newton developed a kind of physics that challenged us in a similar way to move beyond the limits of our common-sense experience by showing us familiar things in a new way.

For example, our everyday experience shows that bodies that are moving always slow down and eventually stop. But Galileo and Newton argued: "Look, an object moving on a slippery surface tends to stay in motion longer. And the more slippery the surface, the longer it stays in motion. Imagine a surface that is so slippery that it has no friction at all. An object in motion on that surface would stay in motion forever!" Conclusion: the natural state of a body is not to slow down and stop, but to move in a straight line at a constant speed. No one has ever actually seen a frictionless surface or a body in perpetual, straight-line motion, but Galileo and Newton started with a familiar experience and took us to and beyond the limits of that experience.

Einstein does something similar in his famous thought experiment in which we ride along on a beam of light; he asks us to imagine how the world would look to us—space and time curved together, rather than strictly separate. That may be mind-bending, but at least there's something recognizable for you to bend your mind around. At least you can picture what's going on, even if you can't do the math.

In that sense, classical physics is like representational art. They both start from our daily experience and then push us to and beyond the limits of daily experience. The basic insights are accessible to anyone with a bit of imagination, even with no background in math (or art history).

But it's a different story when it comes to *nonclassical* physics, like quantum physics. Nonclassical physics strikes lots of people the way that nonrepresentational art strikes me: it leaves them feeling left out in the cold. How do I engage a work of art that seems to just stand on its own without referring to anything in my experience?

Look, we're walking by an example right now: Jackson Pollock's *Greyed Rainbow* from 1953. No figures, no colors. Just an incredibly complicated swirl of white lines on a black background.

GUY: There's something hypnotic about that painting.

PAUL: Maybe, but at least for me there's just no point of entry for understanding it or engaging it. There is nothing familiar represented in the painting. There are just abstract swirls and shapes.

In Picasso's painting, in some ways the old guitarist was unfamiliar: he was blue and angular. But at least he was recognizable as an old man. In Pollock's painting I don't find a similar mix of the familiar and unfamiliar; for me it's all just unfamiliar. Now maybe if I had some training in art, I'd be able to appreciate Pollock's painting. But since I don't have that training—since I'm not an expert—I just find myself feeling alienated by his work.

I think that the discomfort I feel with purely abstract art is akin to the discomfort a lot of people feel with respect to modern, nonclassical physics—and with respect to the Big Bang theory.

Many important quantum concepts have no analogue in daily experience. We can't picture them. In quantum physics, the size, speed, and location of an electron can't be specified all at once. It's not just that we have a problem measuring and determining those things. It's also that, for something as small as an electron, the concepts of "size" and "place" and "speed" no longer mean quite what they mean in our everyday world. That doesn't pose a problem for people who have some training in physics—people who can do the math. But for nonexperts, quantum physics can seem forbidding and alienating.

When you try to connect some of the most basic quantum concepts with everyday experience, you end up in paradox.

GUY: There's the famous example of Schrödinger's cat—Erwin Schrödinger came up with this example in 1935 to illustrate a quantum effect called superposition. Schrödinger proposed a gizmo that operates inside a box with a cat; there's a radioactive atom that either decays or doesn't decay, with some probability—and if it decays, it releases a poison gas, which kills the cat. You won't know until you open the box which event occurred, if the cat is dead or alive. Schrödinger's intent was to point out a strange feature of quantum physics: before you make a "measurement," quantum physics forces you to treat the atom and the cat as if they were somehow in a superposition of two different states, one corresponding to atom decayed and cat dead, and the other corresponding to atom not decayed and cat alive. In quantum physics, our normal idea of a well-defined "state" of an object no longer applies, especially for objects that are very small.

But mostly Schrödinger's example just confuses people—and makes them feel sorry for the cat! Whenever I try teaching with such examples, my students usually get so distracted that they miss the point. You have to live a long time with quantum physics before you can even begin to appreciate how strange it is. The best book I've read that gets across the weirdness of this kind of physics is *The Quantum Enigma* by Bruce Rosenblum and Fred Kuttner, a couple of hippie physics professors at the University of California in Santa Cruz. (And judging from the reviews I have seen online, this book infuriates a lot of people. Some critics argue about how the authors present the role of the "conscious observer" in quantum physics or complain that the authors' descriptions of these effects are "muddled"—a problem with many such popularizations. But others just seem to hate their message—the way that all those assumptions we worked so hard to assimilate in freshman physics are now upended!)

PAUL: As long as the difficult-to-picture effects of nonclassical physics remain safely out of sight, most people can ignore them—sort of like how it's easy to forget about God until you find yourself in a foxhole. But the problem of the strange, abstract nature of modern physics can't be ignored when it comes to the Big Bang theory.

The Big Bang is our best account of the early stages of the universe. It's our best account of our remote past, of our "Deep History." The Big Bang points to a time when the universe was very, very small and very, very dense . . . and therefore to a time when nonclassical quantum effects came into play on the scale of the universe as a whole. This means that, to the extent that science can tell us "Where We Came From," it ends up doing so in terms that seem inaccessible and alienating for those who have not studied physics. I think this may be one reason some people are drawn to scriptural Creation stories instead of the Big Bang theory. At least the scriptural Creation stories are *stories*, which can be pictured and understood.

Imagine for a moment that the whole art world is agog about a new work of art entitled *The Beginning of Everything*. All the professional artists are saying: "This is it! It's a work of genius. It reveals a profound truth about the origin of the universe and about our place in it." You go to the museum to see this great work. It turns out to be a nonrepresentational work of abstract art—all you can see are blobs of orange, green, and blue. You can't make sense of it; you can't enter into it. But all around you, people are oohing and aahing about what a great painting it is and about the deep insight it provides into the origin of the universe. Bitterly disappointed and feeling left out, you leave the museum, resenting those highfalutin artists who presume to portray the "Beginning of Everything" in ways that you can't understand.

That's how some people respond to the Big Bang theory as a way to explain our origins. And their reaction can be intensified when the Big Bang theory is cited either in support of, or against, other origin stories, such as the stories we find in the Bible. The (unfortunate) response of some people is to put stock in the literal truth of the (understandable and accessible) scriptural Creation stories instead of the (alienating and inaccessible) Big Bang theory. After all, if your main goal is to try to come to grips with the meaning and value and purpose of life, you're going to be drawn to an origin story that is understandable and accessible, rather than to one that is alienating and inaccessible.

GUY: It's not only religious fundamentalists who are thus tempted. I was once invited to a Skeptics Society conference, where my host gave a nice, Astronomy 101–level description of the Big Bang. The skeptics in attendance were horrified; they couldn't follow the math and accused him of merely "appealing to authority" when he tried to describe current theories about the first few seconds of the universe.

Even astronomers themselves resisted the Big Bang theory at first. They couldn't "picture it," and it went against their familiar assumptions that the universe was always the same, everywhere and "every-when." They felt more comfortable with a universe that was eternal and unchanging.

The irony is that Lemaître, who developed the Big Bang idea, was not only an astrophysicist but also happened to be a Catholic priest. This may have led some astronomers to suspect his motivation, since the Big Bang might be identified with the moment of Creation described in Genesis. When one atheist cosmologist, Fred Hoyle, mockingly referred to the theory as Father Lemaître's "big bang," the name stuck. (Incidentally, for all their scientific and theological differences, Hoyle and Lemaître became the best of friends.)

Some people did suspect that Lemaître was trying to find a

"scientific basis" for the idea of a Genesis point. Lemaître himself denied this. In 1951, when Pope Pius XII noted the interesting fact that scientists were seriously talking about a beginning point to the universe, Father Lemaître personally urged the Pope *not* to promote his theory as a proof of Genesis. And, in fact, the Pope was careful not to do so. After all, who knows what our cosmology theories will look like in a thousand years?

PAUL: Both atheists and Christian apologists have tried to read theological significance into the Big Bang theory, either as a confirmation of the biblical story of Creation or as a substitute for divine Creation. But it's a mistake (and strategically unsound) to try to prove or disprove religious beliefs on the basis of currently accepted scientific theories, because those theories will likely change someday. As a scientific theory, the Big Bang will always be open to further development; undoubtedly, someday it will evolve into a different and perhaps utterly unrecognizable theory that better fits the data or better answers new questions that we don't even know enough to ask yet.

GUY: That said, it's important to remember two things about the Big Bang.

First, it has been supported by lots of observations since it was first proposed. It's a good scientific theory: it makes testable predictions. And so far, it has passed all the tests.

For example, the expanding energy at the beginning of the universe should eventually turn itself into matter, in accordance with Einstein's famous equation $E=mc^2$. This matter should be in the form of hydrogen and helium atoms, the two simplest forms of matter, in proportions that could be calculated by the theory. And, it turns out, the observed ratio of hydrogen and helium in the universe fits these predictions.

Another condition predicted by this theory is that the leftover radiation from the expansion should be detectable in the background of the universe as radio waves that are nearly isotropic (in other words, the same in every direction) but with a very low energy, equivalent to only a few degrees above absolute zero, as the immense heat of the beginning tiny state has now been spread out over the immense dimensions of the universe in its current state. This microwave background radiation was first observed in 1965.

And the second thing to note about the Big Bang is that, as a theory, it's still incomplete. We now know that there's a lot of stuff we still don't know, stuff we never even would have noticed before, without this theory.

Among the most exciting observations of recent times, looking at the motions of more and more distant galaxy clusters, is that the expansion of the universe is not slowing down, as would be expected as gravity pulls against the initial expansion velocity, but is actually speeding up. Apparently Lemaître and Einstein were both correct: along with Lemaître's initial expansion, Einstein's "cosmological constant" to accelerate the universe also does exist.

Furthermore, various data, including the detailed observations of the acceleration of the expansion of the universe and its apparent lack of spatial curvature, have led cosmologists to infer that, in fact, only 4 percent of the mass in the universe is made of visible matter such as stars and planets. Instead, the essence of the universe appears to be dominated by poorly understood entities called "dark energy" and "dark matter."

PAUL: And if most of the universe is made up of a something-we-know-not-what, which we call dark matter and dark energy, then physics has a long way to go—despite all the progress that has been made. And don't forget that we're still looking for an adequate theory of gravity. According to Newton's Theory of Universal Gravitation, all massive objects (such as planets and apples) exert an attractive force on one another. Einstein's General Theory of Relativity has superseded Newton's theory, but we're still looking for a theory of gravity that would be consistent with quantum physics.

GUY: The recently discovered Higgs boson (yes, the thing that the popular press insists on calling the "God particle") is an important development in comprehending how the concept of "mass" might be understood in the subatomic realm.

For the last hundred years, physicists have been testing their theories about what atoms are made of by smashing atoms together and then keeping track of all the different fragments that show up in their detectors. The Standard Model of particle physics, which has been around for forty years, makes predictions about what sort of particles ought to exist inside an atom, and the Higgs boson was one of the last on the list that hadn't been seen yet.

According to the theory, it would take an especially high-energy collision to shake one of these guys loose, and so a whole new particle accelerator for that purpose was built at CERN, the European high-energy-research lab in Switzerland. Finally in 2012 they detected something that might fit the bill. If, indeed, it's what they were looking for (it'll take years to sort out all the data to be sure), then that would give us great confidence that the Standard Model is a useful guide to understanding what's going on inside an atom.

PAUL: So tell me, Guy . . . in your view, as someone who is a scientist but who has also studied theology and is a person of faith, does anything we know from experiments like these provide a window into understanding whether or how God created the universe?

GUY: In one sense, what we know about the Higgs boson, or the Big Bang, is no more significant than any other observation of nature: every new observation of nature is a window into how the Creator acts.

God reveals himself in the things he has made—that's not me speaking; that's a quote from Saint Paul's Letter to the Romans. (Chapter 1, verse 20, more or less.) Of all the ways God could have made the universe, the fact that he chooses to make it *this* way instead of *that* way tells us something about His "personality."

PAUL: Such as?

GUY: Well, for one thing, stuff doesn't "just happen." The activities we see in the physical universe are not merely the arbitrary whim of some pagan deity. The amazing complexity of the physical world results from rules that are logical and reasonable and, at their basis, remarkably simple.

And while it seems the height of arrogance to think that we humans could possibly understand these rules, we actually do—though always in an incomplete way. Yet we never run out of questions to ponder. The universe, like its Creator, is inexhaustible.

But it's even more than that. The physics behind how a star shines is not only logical and reasonable; it's also beautiful. And so is the star itself. That tells me a lot about Whoever is responsible for those stars!

PAUL: So beauty has a role in science, too, not just in art.

GUY: And, as in art, learning how to recognize that beauty is something each generation has to accomplish for itself. But the joy we get from experiencing that beauty is why the universe is such a delight for us to play in—that kind of "play" we call science. Science is God engaging with us in the best kind of game.

Remember that the Higgs boson was called the "God particle" only metaphorically, to emphasize how important the physicists thought discovering this particle might be, and to suggest that the elegance of a theory that worked with this particle might mirror the elegance of the Creator. (Leon Lederman, who first came up with the phrase, meant it as a joke; until recently we've had to take the existence of the Higgs boson on faith!)

This reminds us that science often works with metaphors. The people who insist on interpreting Scripture "literally" would do well to remember that science itself does not describe things "literally." For example, using the word "particle" to describe the

Higgs boson is a metaphorical stretch. A boson is a tiny subatomic entity that we can really only describe mathematically; it is not recognizably a "particle" in the way that a speck of dust is a particle. Calling the boson a "particle" sets up a mental picture that helps us try to imagine what a boson might be like. But it's a metaphor—a mental representation.

PAUL: Much as the paintings in this museum use the colors of oil and chalk to represent the colors of nature.

GUY: Even the mathematics that we use to describe nature is a metaphorical representation of what nature itself is actually doing. The formula describing the motion of a body as it falls, pulled by the force of gravity acting between its mass and the mass of the Earth, represents only some aspects of the object and its motion. The path of the falling object is *like* the solution to the equations of motion in some senses, but not in others. The accuracy of the calculation depends on how well the formula also accounts for, say, wind resistance or the fact that neither the Earth nor our falling body is a perfectly spherical or uniform mass. And the formulas don't tell us if the falling object is a pearl or a bird dropping.

All representations and metaphors are limited. But it's only through representations and metaphors that we can begin to get a grip on the reality of subatomic particles (or on the reality of love, or on the reality of God). These are cases in which simple human language can't do justice to what we are trying to describe.

PAUL: I remember when I was a kid, we had a great board game called Mouse Trap. Remember how it worked? During the game, players would construct, piece by piece, a ridiculously complex contraption that turned out to be an elaborate mousetrap. I used to love watching the bizarre sequence of mechanical events, all causally connected, that would lead up to a mouse being trapped.

I think that game helped spark my early interest in science and engineering.

Even as a kid I could tell that the mechanism was ridiculous. But I loved the way it all fit together. And it made me wonder whether the world itself was a kind of mechanism—it made me wonder whether all the events I see around me are part of a big cosmic causal chain, a giant cosmic mousetrap. That's how I thought about the Big Bang when I was little: the kickoff move in the cosmic game of Mouse Trap!

Physicists seek to describe the universe as if it were some sort of giant mechanism—a giant mousetrap that behaves in accordance with well-defined causal laws from beginning to end. We hope the universe is not as ridiculously complicated as the game of Mouse Trap; we hope and expect that the mechanisms governing the universe are relatively simple (even if the mathematics used to represent them is challenging and daunting).

But physics does more than just describe the mousetrap, the mechanism at the heart of our universe. Physics also tries to figure out what other sorts of mousetraps might have existed instead. Might the universe have come to be with different physical laws and mechanisms? Or is the universe that we have the only kind of universe that is possible? So physicists don't just seek to explain the cosmic mousetrap that we happen to have; they also try to figure out what other cosmic mousetraps there might have been instead.

What physics *cannot* do, however, is tell us why there is a mousetrap at all: physics cannot tell us why there is something rather than nothing. Nor can physics tell us why we happen to have *this* mousetrap rather than some other one; it cannot tell us why we happen to live in *this* universe with *these* laws of physics, as opposed to some other possible universe.

GUY: There's a mistake some people make when it comes to talking about the Big Bang theory. To use your Mouse Trap metaphor, they want to make God into one of the players in the game, the one who kicks off the first move. But, no, he's more like the guys at the Ideal Toy Company (Marvin Glass, Hank Kramer, and Sid Sackson, according to Wikipedia) who invented the game and made up the rules in the first place!

The Big Bang is a fine description of what happened once things got started, but it can't explain why there are "things" that got started to begin with.

The confusion can be traced back to the scientist-theologians of Newton's day, who saw God as having the job of filling the gaps in physical theory. But even more, the fault rests with Christian apologists who tried to baptize the Prime Mover of Aristotle's physics by identifying it with the Christian God.

PAUL: Some people would like to use the Big Bang theory to eliminate the need for a biblical Creation story or anything like it: "See, science can explain entirely on its own how things came to be. So there's no need for a biblical Creation story. We don't need God to explain why the universe is the way it is; science can explain it all. Or it will, eventually, if we keep working at it."

GUY: As an example, recently two of the brightest physicists working today, Stephen Hawking and Lawrence Krauss, have each tried to answer the question "Why is there something instead of nothing?" by invoking quantum fluctuations in the primordial gravity field that, they suggest, could lead to the Big Bang. It's a controversial theory, scientifically, and it might not be correct.

But even if it is correct, it still won't explain "Why is there something instead of nothing?"—it won't put God out of a job. If the universe came to be from fluctuations in a primordial gravity field, then the question will still be: Why was there a primordial gravity field, and why are there laws of physics that allow it to spontaneously go "bang"?

Besides, if you start out by equating "The Entity That Starts the Universe" with "God," and then you show that gravity is what starts the universe, then you haven't proved there is no God—you've proved that God is gravity!

PAUL: So now we know why Catholics celebrate Mass.

GUY: Grrrrr . . .

PAUL: Bad jokes aside, the same mistake gets made in the opposite direction. Some religious people want to see the Big Bang as providing an argument in favor of the scriptural story of Creation: "See, science has shown there was a Beginning. So the universe hasn't been here forever—at some point it came to be. And that proves the scriptural story of Creation."

GUY: Trying to use "science" to prove the existence of God has the effect of making science a greater authority than religion—it gives science the last word over faith.

It's one thing to say that the complex beauty of nature reflects the glory of the Creator, and to delight in the design we see there; that the reason we do science is to glory in truth and the ultimate Author of truth. That's fine religion. However, to reduce God to simply another force of nature alongside electromagnetism or gravity, as the immediate reason something happens, is bad theology and bad philosophy.

The people who try to do this are usually attempting to rescue a literal reading of Genesis. It's fundamentalism dressed up as science.

And within Catholic circles, there is nothing traditional about that sort of fundamentalism. It is a new and peculiarly modern development from outside the Catholic tradition. As we saw with Augustine, the ancient and medieval theologians were accustomed to interpreting the Bible on many levels, both literal and figurative. But the Protestant Reformation with its emphasis on "the

Word," Scripture alone, had the effect of flattening (they would say purifying) biblical interpretation to only the literal meaning. The eighteenth-century Protestant movement called the Great Awakening cemented this interpretation into the evangelical churches. So biblical literalism is essentially a Protestant idea. It only started to infect some Catholics (ignorant of our history and theology) in the last forty years.

Even back in the 1950s, when I was being educated, the nuns who taught me had no problem at all with modern science. Nor did the Church have any issues with the bigger questions of cosmology. You can find many specific statements by various Popes over the past hundred years and more in defense of modern astronomy and modern science.

Here are a few samples: ". . . the age of spiral nebulae are (about) ten billion years . . . the average age of most ancient minerals [on Earth] is indicated at a maximum of five billion years . . . although these figures are astonishing, nevertheless, even the simplest believer would not take them as unheard of and differing from those derived from the first words of Genesis . . ." That was Pope Pius XII in 1951. "Both religion and science must preserve their autonomy and their distinctiveness. Religion is not founded on science, nor is science an extension of religion. Each should possess its own principles, its pattern of procedures, its diversities of interpretation, and its own conclusions." That was Pope John Paul II, in 1988.

PAUL: In the speech he gave about evolution in 1996, "Truth Cannot Contradict Truth," Pope John Paul II developed an important and ancient Catholic teaching, one that goes back at least to Saint Augustine: the notion of the ultimate "unity of truth."

It goes like this: God is revealed in two distinct ways. First, God is revealed through events of human history, as recorded and reflected in the inspired writings of the Bible and in Church teaching and tradition. But also, God is revealed through the very structure and beauty of nature. This is often expressed in shorthand by saying that God is the author of *Two* Books, the Book of Scripture and the Book of Nature. The traditional Catholic position is that these two books *cannot* disagree with each other, once they have both been properly understood. Both books are written by one and the same author, God. And God does not disagree with God: "Truth Cannot Contradict Truth."

This means that, if there seem to be differences between science and the Bible with respect to the creation of the universe and the creation of life on Earth, that merely means that we haven't yet managed to "read" or "interpret" the Two Books correctly.

So when science and the Bible seem to be in conflict, one possibility is that we don't yet have the science completely right; we haven't yet learned how to read the Book of Nature correctly.

GUY: Science is a trial-and-error affair that takes a long time to arrive at the truth, and I am confident we'll never get to any final "theory of everything" of the sort that would put us all out of business. The more we learn, the more we realize how little we know.

Most of my daily work in science is trying to fit together different bits of data that I am confident are all true but that appear to be at odds with one another. The conflict between science and the Bible is nothing compared to the conflicts you can run into between science and science! In fact, that's the most fun you can have as a scientist, because when you see apparent contradictions in your data, you realize you're about to learn something new.

So even if we're confident that science is headed toward the truth, we're not there yet. And we're not yet sure which parts of our science are true and which parts aren't true. So we're never sure to what extent we are reading or interpreting the Book of Nature correctly.

PAUL: The other possibility, when science seems to be at odds with the Bible, is that we haven't yet interpreted the Bible correctly; we haven't yet learned how to read the Book of Scripture correctly.

Maybe we're reading the Bible literally, when we should be reading it figuratively—or vice versa.

The third possibility is that we're reading *both* books incorrectly—we have both the science and the scriptural interpretation wrong!

The Catholic tradition is that the proper response, when the Bible seems to be in conflict with science, is to have *faith in the ultimate unity of truth*. We can live with apparent conflicts between science and the Bible, at least for now, because of our faith that, in the end, there will be no conflict between them. We can't see that now, not yet—because we can't yet read the Two Books correctly. The upshot of this Catholic faith in the ultimate unity of truth is that we don't need to insist that science and Scripture agree in all respects *right now*. We can keep on doing good scientific work, trying to understand and interpret the world. And we can keep on studying the Bible, trying to understand and interpret God's Word better.

GUY: For one thing, such conflicts between religion's and science's description of the natural universe rarely, if ever, come up in practice. The conflicts you read about in the popular press are usually not about science, but about the use of science. No one doubts the biology behind stem cells; the issue is not whether the science is accurate, but whether using the technology based on that science is a good idea. (Sadly, we've learned that even when people have the best of intentions, all too often applying new technologies will result in unintended consequences. The risks aren't always immediately obvious to the guys who come up with the bright ideas.)

In other words, religion worries about not just what's happening physically, but who's doing it, and why, and what are the possible side effects to individuals or society—the very stuff that modern science deliberately removed from Aristotle's old way of doing science.

PAUL: But even if there should somehow arise an apparent contradiction between what we believe in faith and what science tells us about how the universe works, we don't need to panic—we can hold that conflict in tension. Why? Because of our faith in the ultimate unity of truth. We have faith that, in the end, when we understand both the world and the Bible perfectly, we'll see that science and religion are not in conflict with each other. One and the same God wrote both books, of Scripture and of Nature, and God will not disagree with Himself: "Truth Cannot Contradict Truth." Because of this faith, we're able to live with any seeming conflicts between science and religion.

GUY: The great irony in all of this, as far as I am concerned, is that it's the religious fundamentalists who end up being the ones who seem to be lacking in faith! When they insist that science and faith must be in agreement *right now* (and that, therefore, science should yield to faith), it shows that they don't have faith in the ultimate unity of truth . . . they don't have faith that God is the author of both books, of Nature and of Scripture . . . and they don't have faith that God has made us in His image, with a limited-but-real capacity to read and understand both of the Two Books.

PAUL: I pray for fundamentalists on this point. I pray that they will grow deeper in a kind of faith that will permit them to let go of the destructive insistence that science must agree with the Bible in all respects *right now.*

But it's not only religious fundamentalists who have this sort of difficulty. There are also scientific fundamentalists. People like Richard Dawkins take the fact that a literal interpretation of the Bible sometimes disagrees with current science as an indication that the Bible should be dismissed altogether. Just as religious fundamentalists want to put science under the judgment of faith, scientific fundamentalists want to put faith under the judgment of science. Both kinds of fundamentalism seem to be driven by a fearful need to have utter certainty and consistency.

GUY: And to have it *right now*.

I like that idea of "science fundamentalists." After all, what is a "fundamentalist" but someone who has a flattened, one-dimensional view of the subject, and who thinks that if his view of the universe is true, then necessarily all other views must be false? When you think that the "fundamentals" are all that's important, or that knowing the "fundamentals" is enough by itself, it's like seeing only the dots of paint in Seurat's painting.

PAUL: And, what is worse, it leads you to try to use the "fundamentals" of one topic to solve the issues of a different topic. You're using the wrong tools to try to answer questions they were never designed for. It's misguided to ask whether the Big Bang theory would provide evidence for or against a scriptural story of Creation; the Big Bang Theory is a scientific theory, concerned with physical causes that are proximate and contingent. And it's wrong to ask whether a scriptural story of Creation would provide evidence for or against the Big Bang theory; scriptural stories of Creation are concerned with ultimate origins and with humanity's personal relationship with God. Those are the wrong questions for the tools at hand.

GUY: But if those are the wrong questions to ask, what are the right questions to ask? What can I say to people who ask me about connections between the Big Bang and scriptural stories of Creation? It doesn't seem right simply to say, "Forget it—that's the wrong question to ask!"

PAUL: The Big Bang theory and scriptural stories of Creation both seek to answer the question, *Why are we here?* But that question means something completely different to a scientist than to a philosopher.

Science only gives *proximate* answers; each answer leads to a new question. Do you want to know why objects fall more slowly on the moon than on Earth? Isaac Newton's Theory of Universal Gravitation will tell you why. Do you want to know why Newton's theories are true? The Standard Model of particle physics will tell you why. Do you want to know why the Standard Model of particle physics is true? Maybe someday a scientist will propose a theory that answers that question, in terms of something more fundamental . . . and so on, down the line, to broader and broader explanations. But no matter what newer, broader explanation gets proposed, you can always come back and ask: *Why is that true?*

Once when I was on vacation with my family, my little nephew Craig got off on an unending series of "why" questions that drove me nearly out of my mind. If we hadn't been on vacation, I would have told the kid to shut up—or I would have run away! But since we were on vacation, and I had nothing better to do, I kept answering his questions, just to see how far it would go. Craig started with *Why are there waves on the ocean?* That led to *Why is there wind, why are there clouds, why is the sky blue, why is sand small, why are seashells sharp, why do fish smell funny* . . . and so on. There was no end to it. Every time I explained A in terms of B, Craig asked me to explain B. And that is just what scientists do. (Now my nephew Craig has a son of his own. Oh, sweet revenge! I can't wait till little Jack starts plaguing Daddy Craig with endless questions. I want a front-row seat for that one.)

So in a sense, scientists are like toddlers, with their enormous capacity to ask *Why*. But in another sense, scientists are unlike toddlers. When toddlers ask *Why*, they are still trying to figure out whether the world hangs together—they are still trying to figure out whether the world makes sense at all. That is something they have not yet learned—or, rather, something they have not yet learned to take for granted. When toddlers ask *Why*, they are still experimenting with what it *means* to ask why.

Not so with scientists, who assume—who take for granted—that nature hangs together and makes sense. Scientists (unlike toddlers) presuppose that nature is held together by causal laws and can be understood by human reason. So scientists know what they mean when they ask *Why*.

Does nature really make sense? Is nature really susceptible to human understanding? Does nature really hang together by means of causal laws? The answer to those questions must be either yes or no: there must be a fact of the matter. Either the universe is rational, or it isn't—either it is a rational *logos* or an *a*rational *chaos*.

GUY: *Logos* comes from the Greek word for "word," and it's the root of our English word "logic." (Not to be confused with LEGOs, which are a different kind of building block!) It's also the word used at the beginning of John's Gospel, "In the Beginning was the Word . . ." Christ is "the Word," the *logos*; thus, he is identified with logic, reason, and order. Those are the things that will redeem a world that can often seem chaotic.

PAUL: But here's the kicker. Merely by posing the question of whether the universe is a *logos* or a *chaos*, you are already assuming that it is a *logos*. If you are asking a *Why* question, and you expect an answer, you are already assuming that the universe is somehow rational and intelligible—that it is a *logos*. If the universe were a *chaos*—if it were not susceptible to being understood by reason—then a *Why* question would have no meaning at all. So if you even ask the question, Is the universe a *logos*? you are already presuming that the answer is yes.

GUY: And just because you can find examples in the universe of what we call *chaos*, that doesn't mean the whole universe is chaotic.

A roulette wheel or a pair of dice gives you "random" results only as the result of careful design and construction. And while there are mathematical systems we call chaotic, there is, in fact, a deep mathematical structure underlying them.

I am reminded once more of the abstract art we've looked at in this museum. Even if the splattered bits of paint were applied randomly to the canvas, they were done so by an artist who chose to do so, and who then chose to frame and exhibit this result and not some other version, for a reason.

PAUL: Science cannot answer the question: "Why is the universe a rational *logos*, rather than an *arational chaos*?" That'd be like using science itself to answer the question, "Why is science possible?" Or it'd be like trying to use reason itself to answer the question, "What are the grounds of reason?" The grounds of reason cannot be explained by reason itself—if they could, then they wouldn't be grounds of reason! So we can't expect a reasonable explanation if we ask, "*Why* is the universe rational?" We can't expect a reasonable answer if we ask, "*Why* can we ask *Why*?" In these cases we are like toddlers: we don't really know what it is we mean when we ask *Why*. The best we can do is to keep asking the question in different ways, as my nephew Craig kept asking me questions. We hope that, by testing the waters with many questions, eventually we'll figure out what it is that we are trying to ask.

That is part of what is going on in the various Creation stories in the Bible. The Creation stories keep posing the same ultimate questions: Why is the world intelligible and rational? Why are we here—what is our part?

The biblical Creation stories suggest that the kind of answer to be sought to these questions is connected somehow with goodness and with love. When God created the world, God "saw that it was good." And God—God who is the grounds for the being of the world—is Love.

You don't need to believe these things to be able to do science.

You can do good science whether or not you think the world is good, and you can come up with the Big Bang theory whether or not you think that God is Love. But the goodness of the world and the love of God are where Christians locate the answer to the question *"Why can we ask Why?"*

GUY: When I am asked, "How do we understand the Big Bang in light of Scripture?" it's like asking me how I understand that painting of the blue guitarist "in light of Scripture." What does Scripture have to do with it? Maybe nothing; maybe something very deep. I could come up with an answer that was silly or one that was profound. But no sensible answer would ever try to equate the painting with Scripture.

The painting shows me something about the guitarist I never would have noticed on my own. But, of course, there's more to the guitarist than what's on the canvas. In the same way, the Big Bang is a scientific painting to describe what happened at the beginning of the universe, just one of many such pictures that have been painted over history. (And it's a subject that will continue to be painted well into the future; the theme has hardly been exhausted.) It tells us things we might never have realized about our origins and our history.

But the Big Bang theory in itself is hardly itself the sum total of the universe. I can't find ultimate meaning from that theory, any more than I could get a piece of pie from the diner in Hopper's painting *Nighthawks*.

Speaking of which . . . isn't there a coffee shop in the basement?

Day 2: What Happened to Poor Pluto?

SETTING: EAST ANTARCTIC ICE PLATEAU

SOLES ON ICE

GUY: Since we traveled to Chicago yesterday, I thought we might go farther afield today . . . at least in our imagination. (I am of the generation who grew up believing that "thinking is the best way to travel." In reality . . . to travel from Chicago, to New Zealand, to McMurdo Station, to the inner plains of Antarctica? With all the typical weather delays, we'd be lucky to get there in a week.)

Why Antarctica? Well, we're talking a lot about science and religion in these conversations, but the fact is that while most people have experienced religion, for better or worse, fewer people actually know what science is really like from the inside. So I thought, why not go someplace that's empty of almost everything *except* science? There are few places on Earth as empty as the East Antarctic Ice Plateau.

And the example I want to use for how science works concerns a place even colder than Antarctica, one that was in the news a lot not so long ago: Pluto.

PAUL: The Pluto story points to some of the basic ways that science is done—how the scientists themselves behave, especially when they don't realize what they are doing while they're doing it. It also raises the question of how we understand change in science—what happens when what we take to be true changes?

And that has implications for how we relate scientific truths to religious truths.

But I have to confess . . . I've been feeling bad for poor Pluto ever since you astronomers decided to demote it. Now Pluto is no longer called a "planet." Instead it's called a "dwarf planet." I realize that Pluto itself probably doesn't have feelings about this. But speaking for myself and for lots of other people who grew up memorizing the names of the *nine* planets, this comes as a bit of a blow.

GUY: You're not the only one who has hurt feelings on this issue. About a year after Pluto had been reclassified, I was giving a talk about it at the Cranbrook Institute of Science. It's located in the northern suburbs of Detroit, where I grew up—in fact, I used to ride my bike there all the time when I was a kid. It was great fun going back there as a scientist, and it was a way for me to pay them back for all the good they had done for me.

So, anyway, my talk went into the kinds of observations I make with the Vatican's telescope in Tucson, Arizona, on Trans-Neptunian Objects (TNOs) orbiting alongside Pluto in the far reaches of the solar system. I explained the essentials of planetary nomenclature, full of typical computer graphics and erudite astronomy. But then, after the talk, this eight-year-old kid who reminded me of myself at that age came up to me. It was clear all my explanations had gone right over his head. All he did was burst out: "B-b-but . . . what happened to Pluto?"

I felt for the kid. I'm glad that he cares a lot about science, and I hope that maybe he'll grow up to be a scientist. Though it was difficult to explain this to him, really *nothing* has happened to Pluto. But a lot has happened to us astronomers and our understanding of what Pluto is. The "demotion" of Pluto was a reflection and result of those changes.

PAUL: Maybe the question of the status of Pluto, whether it's a planet or not, should only be a matter of "inside baseball" among you astronomers—stuff that is of interest only to insiders. Maybe the rest of us should just leave you alone and let you do your work. We should let you call Pluto whatever you want. After all, we wouldn't get all worked up if biochemists decided that some particular substance will no longer be counted as an enzyme, or if botanists decided to reclassify some species of flower. Stuff like that happens all the time.

But I, too, have sympathy for that eight-year-old kid. The idea of Pluto being a planet is something I grew up with—it's attached to childhood memories, and, thus, it's part of who I am. This sounds strange to say, and it's not particularly rational, but when you astronomers announced that Pluto is no longer a planet, I felt kind of violated. It was as though you'd taken something precious from me, something from my childhood. My first, gut-level response was: So Pluto is no longer a planet? What's next? Will you announce that Mickey is no longer a mouse? That Donald is no longer a duck? What is wrong with you people?

The whole affair reminds me of the upset in my parish, back when I was a little kid, when the Catholic Church went through a lot of changes after Vatican II. Those changes affected the lives of regular Catholics in ways that mattered to them; in particular, the changes in the way the Mass was celebrated. Some people were happy about the changes coming come down from Rome after Vatican II, which seemed to them to be long overdue. But not everybody. The Mass, and the way it had been celebrated always with the same language and gestures everywhere in the world, had been an important point of stability in their lives of faith. To them, changing the Mass meant losing something vital and important. It meant letting go of something they had taken to be true-for-all-time. That made them sad and angry, and it caused some tensions and divisions in our parish for a while.

It's interesting to me to see that something similar can happen when things change in science. Little kids who grew up after Vatican II never had the experience of the old-style Catholic Mass, so they can't really miss it. And little kids who grow up after the reclassification of Pluto will never have known Pluto as a planet, so they won't really "miss it," either. But for those who are older, who went through the changes, there will always be an emotional twinge associated with what was gained and what was lost—when the Mass was changed and when Pluto was changed.

I fully get that science shouldn't be constrained by nonscientific factors like "emotional twinges." If science starts trying to work around nonscientific factors of that sort, it'll never advance. But the emotional twinges indicate that the larger public cares about science, even if not always for the right reasons. And they indicate that the results of science are entangled and enmeshed in everyday life.

So we have to expect that when science advances, it will sometimes meet resistance from the nonscientific public—perhaps strong resistance. This is because the nonscientific public is sometimes heavily invested at an emotional level in this or that aspect of science. This means that science can't be pursued purely as "inside baseball."

Speaking of "inside," can we go inside? My feet are freezing! I thought that Chicago in the winter was cold. But the cold in Chicago is nothing compared to the cold here in Antarctica. Yeesh!

GUY: I told you to dress warmly. That's why they issued us those big white boots. I put some chemical foot-warmers in mine—and in my gloves, too, for that matter.

But I can't think of a better place for the two of us to talk about Pluto. Here in Antarctica, out on the ice, is as close as we can come on Earth to what it must be like to stand on another planet.

PAUL: Or dwarf planet.

GUY: There's a wonderful scene in one of my favorite science-fiction novels, Robert Heinlein's *Have Space Suit, Will Travel*, where his heroes are trapped on Pluto, and he does a great job of describing what it's like to be in such a remote, cold place:

> The sun . . . looked no bigger than Venus or Jupiter from Earth . . . The ground was covered with snow, glaringly white even under that pinpoint sun . . . What sort of "water" was that? Methane? What was the "snow"? Solid ammonia? . . . A wind blew from our left and was not only freezing that side of me, it made the footing hazardous . . . Would a man struggle before he shattered himself and his suit, or would he die as he hit? . . . The wind not only frightened me, it hurt. It was a cold so intense that it felt like flame. It burned and blasted, then numbed.

You can taste the cold in his description.

Of course, writing in 1958, Heinlein grossly overestimated Pluto's gravity and the density of its atmosphere. His guess about methane being present was spot-on; but instead of ammonia—a nitrogen-hydrogen compound—the nitrogen on Pluto is the same as the pure nitrogen in our own atmosphere . . . only frozen into ice.

And don't forget, Pluto is a lot colder than here in Antarctica. We're here at the height of summer; it's only about twenty below today. Of course, the wind makes it feel more like forty below.

PAUL: Fahrenheit or Celsius?

GUY: Ha! Trick question. Can't catch me on that one.

PAUL: Right, forty below is the one temperature that's the same value in both scales.

GUY: The old hands here in Antarctica told us that you can tell when the temperature hits forty below: when you take a pee, it freezes before it hits the ground. Or should I say, before it hits the ice? I learned a lot of useful trivia like that back in 1996. That's when I spent six weeks living in a tent here on the ice, hunting meteorites with the Antarctic Search for Meteorites team. Even though we were here during the Antarctic spring, November and December, it was cold. A couple of days, the wind chill went to seventy below.

PAUL: I read all about your stay in Antarctica in your book *Brother Astronomer: Adventures of a Vatican Scientist*. And, of course, you and I have talked about it with each other often enough. That's why I wanted to come here with you. I wanted to get a taste of what it was like for you and your fellow researchers here in the cold and ice for all those weeks.

But why is it that you came to Antarctica to search for meteorites? Is it that more meteorites actually land here? Or is it more like the case of the guy who looks for his lost keys under the streetlight, not because they're more likely to be there but because there's more light?

GUY: In one sense, it is exactly like the guy looking for his lost keys.

Meteorites fall from space onto the Earth at exactly the same rate everywhere, no matter where you look. There aren't any more falling on Antarctica than on anyplace else. But there are three advantages to hunting for them in Antarctica.

First, like the car keys under the lamppost, they're easier to find here—with the added advantage that, unlike the car keys, the meteorites actually are here to be found. I mean, most meteorites just look like an ordinary rock from a distance, and when they land on a mountainside, it's almost impossible to spot them among all the ordinary Earth rocks. But on the ice? Meteorites are black, and the ice is white; they jump out at you.

Second, most meteorites are filled with little flecks of metallic iron, which will start rusting as soon as they hit our atmosphere. Something that lands in a jungle will have rusted into dust in maybe twenty years. But here, they stay frozen solid for thousands of years. So they can sit around for a long time, waiting for us to collect them; thus, there are more to collect.

And finally, it turns out that the ice cap here slowly moves with time, creeping down toward the oceans, carrying the meteorites with it. But if the ice reaches a spot like this, near a mountain ridge, it stops moving and starts to pile up. Eventually the dry winds sublimate the ice away; it evaporates, but the meteorites are left behind. So the motion of the ice sheet acts like a big conveyor belt, gathering up meteorites from all across the continent and piling them up in regions of blue ice, like this one.

Besides all those sensible reasons . . . coming to a place like this is a blast! A cold blast, but a blast nonetheless.

PAUL: Meteorites must be really important to you, if you were willing to put up with this climate to search for them. I can't imagine how you folks managed to survive.

So, what about this little black rock that I just found here on the snow? Is it a meteorite?

GUY: Don't touch it!

PAUL: Will it contaminate me? Am I going to pick up some sort of deadly space virus?

GUY: Of course not. I don't care about that. I'm afraid *you* will contaminate *it*!

PAUL: Ah, like the crime scene on some TV cop show—I get it. That's another reason we're wearing these gloves, I guess. OK, I'll leave well enough alone. But take a quick look at it. Can you tell me anything about it? Is it special in any way?

GUY: Sure. Actually, talking about this sample is a nice example of how science works.

The first thing you want to do with a rock like this is to try to determine what sort of meteorite it is—which is to say, what other samples is it like? How is it different from other samples? In other words, classification.

You can't tell for sure simply by looking with the naked eye, but there are ways we can guess which type this guy might be. Some meteorites are lumps of metallic iron and nickel, but they're rare; those are obvious—they're heavy and dense. Not this one. Other rare ones are basalts, chips of lava from some body that was big enough to melt and recrystallize. Those are the best; they might be from the moon or Mars. But their black fusion crusts—formed as they scream through Earth's atmosphere, burning off their outer layers just before they hit—have a certain distinctive shine, which that one doesn't have.

PAUL: So the first criteria you use to classify a sample in the field are density and visual appearance.

GUY: Most meteorites, like this one, are what we call "ordinary chondrites." "Chondrites" means they have little millimeter-sized beads of molten, crystallized rock called "chondrules," and they're "ordinary" because . . . well, they're the most common type we find. They make up about 80 percent of all the meteorites we recover here in Antarctica.

PAUL: And the only way you can make a statement about which types are common and which are rare is if you have a lot of samples to study and a well-defined way of sorting these samples into different types. Of course, the more samples you have to classify, the more people you want to have doing this work; there are too many samples for one person to do it all. So an essential first step to your work is to have a way to classify different meteorites that everyone can agree on.

GUY: Right. And then, once you have sorted them into groups, they can be further subdivided into various types, like H or L if they are high or low in iron, or C if they are carbonaceous, or E for enstatite-rich chondrites . . .

PAUL: So you develop your own specialized terminology for classification. For instance, what is "enstatite"?

GUY: A variety of the mineral pyroxene, made from silica and magnesium oxide but almost no iron oxide; in these meteorites, the iron is all in the metallic form. It's . . .

PAUL: The point is, you specialists classify your meteorites into different categories or kinds based on their chemical and mineral composition. And for each class you have particular criteria, and specific names, that everyone agrees on.

GUY: Of course. And then they're subdivided by their physical state, how much they've been shocked or heated.

Sorting and classifying is the first thing you do with any collection of data. That allows you to look for patterns among the data. Then, once you've found the patterns, you can start trying to understand what makes those patterns happen. So classifying meteorites is essential.

PAUL: But I wonder: Sometimes, does the terminology need to be changed as we learn more and more about an object? Have the categories that you use changed over time—have you ever had to reclassify meteorites according to new categories, just as Pluto has been reclassified?

GUY: You bet. For example, there was a big shift in the 1970s in how carbonaceous chondrites were named; what used to be labeled with numbers are now labeled with letters. To make life more confusing, some of the samples we used to call C1 are now called CI, but the ones that were C0 (the number zero) are completely different from the ones now called CO (the letter O). It was years before the shift finally took hold in the meteorite community.

What often happens, though, is that we wind up using new definitions for old names, which gets very confusing. The "carbonaceous" meteorites were called that because they are very dark, like carbon, and, indeed, some of the first ones measured were rich in carbon. But it turns out that most of the ones lumped together in that class (for other chemical reasons) actually don't have all that much carbon; they just happen to be rich in other dark minerals, such as magnetite. So they're sorted now into separate classes and subclasses, some rich in carbon, others not. We still call them all "carbonaceous," but the term has lost its original meaning.

Or, another example from closer to home . . . The subclasses of enstatite-rich chondrites had been labeled EL and EH, because people thought they were "Low" or "High" in their metallic iron content. They are certainly distinct classes for other reasons; but when my lab-mate, Brother Robert Macke, SJ, measured their densities and magnetic properties in our lab at the Vatican Observatory, he found that, on average, there's no difference in the metal contents between the two groups. These groups do have important differences, but they have nothing to do with being "high" or "low" in metal. We keep the old terminology just because it would be too confusing to change it all at this point.

PAUL: So, sometimes the basic categories that are used in science change, and, as a result, sometimes things have to be reclassified. That has happened before, even with planets. The controversy about Pluto wasn't the first time people have locked horns as to whether or not a body should be considered a planet.

GUY: Well, yeah. Remember what we described yesterday in the Art Institute? For the ancient Greeks, a *planet* was a "wandering star"— something that looked like a star but that moved from month to month relative to the rest of the stars, which remained fixed. The Greeks considered the sun and moon to be planets, along with Mercury, Venus, Mars, Jupiter, and Saturn. But for them, the Earth did not count as a planet, since it didn't move—it was at rest at the center of the universe. The Greeks also noticed the occasional *comet*, which moved relative to the stars. But comets moved relative to the stars in ways that were irregular and unpredictable, while the motions of planets were regular and predictable.

PAUL: But my point is that, during the 1600s and 1700s, the definition of *planet* itself went through a big change. Following the widespread acceptance of Copernicanism (heliocentrism, the idea that the planets move around the sun, not the Earth), a *planet* came to be understood as an object that moved in a regular orbit around the sun. So the sun itself was no longer considered a planet—it got bumped from the list. And Earth was added to the list of planets, since it was now understood to move regularly around the sun, along with Mercury, Venus, Mars, Jupiter, and Saturn. At the same time, the moon was removed from the list of planets, becoming the charter member of a whole new category, *moons*—objects moving in a regular orbit around a planet. Our moon became just one of a class, *a* moon, along with the four moons of Jupiter, which were first observed by Galileo.

In subsequent centuries, additional planets were discovered. First to be found was Uranus; and then Ceres and Vesta and a few other small bodies orbiting between Mars and Jupiter; then Neptune; and, later, Pluto. Many additional moons were found, as well. But in the mid-1800s, after a raft of tiny objects were found orbiting alongside Ceres and Vesta, they were all (including Ceres and Vesta themselves) renamed "minor planets," or "asteroids," and removed from the list of planets.

GUY: History repeats. New observational data in the past led scientists to redefine "planet." And that's what happened again recently, with the result that Pluto's status as a planet has been changed.

SEARCHING FOR NEW TRUTH

PAUL: So, what's the history of these expeditions to Antarctica? Who had the bright idea to come here to look for meteorites? How long has this been going on?

GUY: The ANSMET program—the Antarctic Search for Meteorites—had its first season here in 1976. The first director of ANSMET, Bill Cassidy, wrote a book about the history of the program: *Meteorites, Ice, and Antarctica.* I remember when I was just a young postdoctoral fellow, hearing a talk about that program by Ursula Marvin, a geologist at Harvard. She was one of the first to look for meteorites here, in the late 1970s. I've wanted to come ever since.

But actually, it was the Japanese who started the systematic searches here first. Keizo Yanai, who once hosted me on a visit to Japan, was on a Japanese Antarctic survey in the early 1970s that started collecting these samples. (He was also a part of the first ANSMET team, to pass on what he'd learned.) The Japanese program is quite different from ours. We fly down here from Christchurch, New Zealand, on military planes and spend six weeks living in tents on the ice. They come down by ship from Japan as part of a larger Antarctic research program and live in more permanent camps here for eighteen months!

I explained before why Antarctica is an ideal spot to look for meteorites; however, hot, dry deserts such as the Sahara are also places where we find a lot of samples. In fact, the famous French author and aviator Antoine de Saint-Exupéry—the guy who wrote *The Little Prince*—describes in another one of his books,

Night Flight, landing in the Sahara and recognizing the black rocks on the sand around him as meteorites.

Up until these expeditions, we mostly found meteorites either by accident or by immediately hunting for the pieces seen to fall in a bright fireball. The thought that we could harvest samples systematically in a place like this was a new idea. Having access to such collections has changed both the way we think about these samples and the way we do science with them. For example, iron meteorites are common in museum collections because they are easy to spot in some farmer's field; however, with the unbiased sample of all the different types of meteorites we find on the ice here, we now have a better idea of how truly rare iron meteorites are compared to other meteorites.

PAUL: And that gets back to what we said before, about the central role of classification in science and the need for a common way to do classifications. Which I guess was what was going on with Pluto when it got demoted—am I right? Was it as controversial among the scientists as it was among the general public?

GUY: The news about Pluto's demise drew a lot of comment, and certainly there was some controversy among scientists. But while the reaction was especially intense among the general public, in one sense it's a bit of a tempest in a teapot—the stakes are fairly low. And because the stakes are low, it makes an interesting test case of interaction between science and the larger culture. When the stakes are low, we have a better chance of looking at that interaction in a way that's reasonable and objective! (As opposed to how difficult it is to discuss that interaction calmly and rationally when the stakes are high, such as in the case of Galileo. But we'll come to Galileo in a later chapter.)

PAUL: So, who was it who decided that Pluto is no longer a planet?

GUY: That was the International Astronomical Union, better known as the IAU.

PAUL: Who are they? And who gave them the right to make decisions like this?

GUY: "They" are "us"—the astronomers of the world.

The IAU was founded in 1919, soon after World War I, with representatives of the national astronomical associations from all over the world. The idea was to promote international cooperation in astronomy. One important job the IAU took on was to set up international standards that all astronomers could refer to: arbitrary but necessary definitions, such as the boundaries of constellations, or the names of newly discovered moons and asteroids. A lot of them are technical definitions about coordinate systems and the like, which would take a whole book to explain.

It's thankless work, done mostly by volunteers sitting on committees and working groups. Then every three years the IAU gets together in a General Assembly, where all major decisions get ratified, or amended, by a vote of the nations that have astronomers attending the meeting.

The idea of having nations represented at this meeting, rather than letting the decisions be made by whatever astronomers decide to show up at the General Assembly, was a leftover from the turbulent times of nationalism when the IAU was formed. But it serves a very practical function even today, because it makes sure that no one country dominates. Just because one nation is bigger or richer doesn't mean that it should be able to dominate the discussions. Even small countries get a voice. Small countries . . . like the Vatican City State. That's how I, an American astronomer working at the Vatican Observatory, wound up getting involved in the deliberations about Pluto.

PAUL: So, the Vatican actually was involved in the decision about Pluto?

GUY: It's true. In fact, there were two Vatican astronomers who figured prominently in the decision. I was one of them; Father Christopher Corbally, SJ, the British Jesuit who studies stellar spectra with our telescope in Arizona, was the other. It all came about because the Vatican City State is a national member of the IAU, and, as the astronomers at the Vatican, we represent the nation where we work, not where we happen to be citizens.

My part in the story goes back to the IAU's 1994 General Assembly, held in the Netherlands. The members of IAU Commission 16 (which deals with planets and their satellites, or moons) were organizing who would sit on its governing board for the next three years. They trotted out a list of "usual suspects"—the grand old men of the field (indeed, mostly men back then)—until someone noticed that everyone on the list was either American or Russian.

That wasn't surprising, because in those days no one else was sending up space probes. Nowadays, of course, the Europeans and Japanese are quite active, not to mention recent missions from India and China. I was there, though, representing the Vatican City State. A friend sitting next to me turned and said, "You're a European!" Never mind that I was born and raised in Detroit; my Vatican status filled their quota, and they put me on the commission panel.

And so, nine years later, when my turn came up to be president of the commission, I was named, *ex officio,* to be a member of a nineteen-person working group that was supposed to look into how to deal with Pluto and all the other large objects being discovered out beyond Neptune.

But meanwhile, our colleague Father Corbally was in a position where he would play an even more central role in the decision. He was the chair of the Resolutions Committee who actually had to write the final definition that came up for a vote.

PAUL: But why does it matter to you whether Pluto is called a planet or not? Are there really practical implications for you astronomers or the way you do your work?

GUY: When we said Pluto is not a planet, what it meant for us in the practical sense is that we will keep the data for its orbit and other characteristics in the same tables as the data for other small bodies and no longer keep it with the data for the major planets. When we determine the names of new bodies that look like Pluto, we won't use the same rules used for naming planets (which is just as well, as there aren't any such rules); instead, we'll use the rules and the committees already set up for naming other bodies in the solar system.

It's a very pragmatic decision. For example, big planets are in pretty darn stable orbits, and so you don't have to keep track of so many terms to be able to find them in space. Smaller bodies, such as asteroids, have their orbits pulled out of shape all the time by the tug of gravity from the planets, so you have to keep track of a different set of data in order to be able to calculate with accuracy where they are likely to be found at some future date. Pluto is of a size that it can be (and is) strongly perturbed by Neptune; it simply makes more sense to keep its orbital data in the same database as the other small solar-system objects.

Now, you can go ahead and call Pluto a planet, if you want. Heck, you can call it an orange if you want—I can't stop you. But if you want to be able to talk about Pluto with all the other scientists who are working on it—for example, if you want to know where to look for the tables that contain essential data for Pluto alongside the data for other similar bodies—it makes sense to use the same terminology as the people who maintain those tables. If you look for Pluto under "Oranges," you're going to have a hard time finding anything useful.

Science isn't a big book of facts. It's a conversation. Conversations need a common language; this was a case of deciding on a convention that made the most sense for the scientists who are participating in the conversation.

PAUL: So, where did the confusion about Pluto start—and why did it come to a boil only recently?

GUY: The whole problem started with the bane of every scientific hypothesis: the unexpected coincidence.

In 1846, two celestial dynamicists, John Couch Adams and Urbain Le Verrier, seeing how the orbit of Uranus (discovered only fifty years earlier) had been slightly perturbed, had predicted that there should be a previously unknown planet doing the perturbing. They even predicted where to look for that planet. When Johann Gottfried Galle looked where they'd indicated, he discovered Neptune. The next obvious question was, could you play the same game to find yet another planet out beyond Neptune?

Remember, big planets don't shift their orbits very much, even in the presence of other big planets. The effect these men were looking for would be very small. Nonetheless, by the end of the nineteenth century, some astronomers—including a wealthy gentleman-scholar named Percival Lowell—felt that enough was finally known of Neptune's orbit to be able to narrow down where you should look for yet another planet.

Now, Lowell had made a name for himself (especially in the popular press) by claiming to see canals on Mars, so some of the more "serious astronomers" tended to be skeptical of his work. On the other hand, he was independently wealthy and had built a really nice observatory outside of Flagstaff, Arizona, so he was able to employ astronomers who did have a solid reputation in their fields.

Lowell calculated that the planet he suspected was perturbing Neptune would be ten times the mass of the Earth, in a particular orbit, and he set his observatory staff to search the sky where he predicted it should be found. But for twenty years the search was unsuccessful. Then, in 1930, fifteen years after Lowell's death,

Clyde Tombaugh finally did discover something orbiting out beyond Neptune, pretty much where Lowell had originally predicted. Oddly, though, it was much fainter than anyone had expected.

The director of the Lowell Observatory, Vesto Slipher—not Tombaugh, the planet's discoverer—chose the name Pluto. He announced it in letters to the American Astronomical Society, the Royal Astronomical Society, and the *New York Times*. (Note that the IAU was not involved at all.)

Pluto was sufficiently far away that no one was surprised when even the best telescopes of the day could not resolve it into a disk; no matter how big the telescope, it was never more than a point of light. Given that it was surprisingly dim, even if it was (presumably) ten times as massive as Earth, astronomers assumed it was just small and very dense.

But a funny thing began to become apparent . . . They knew that you could measure Pluto's actual size by watching it pass in front of a background star—the event is called an "occultation"—and measure how long it took for the background star to blink off and then back on: the longer the blink, the bigger the planet. But Pluto kept *missing* the stars it was supposed to occult! Each event seemed to indicate that Pluto was much, much smaller than previously predicted. Soon it became clear that, in order to have the mass that Lowell had calculated was needed to perturb Neptune, Pluto would have to be unreasonably dense, indeed.

Then, in the 1970s, two discoveries finally killed the idea that Pluto was a Neptune-shaking planet. A new analysis of Neptune's orbit, with nearly a century's more data, showed that the tiny "perturbations" used by Lowell and others to predict Pluto were actually just errors and uncertainty in mapping Neptune's position. By choosing which data set to include or drop from the analysis, you could derive any mass you wanted for Pluto—even a negative mass! The prediction that Tombaugh had used to discover Pluto was based on nothing but smoke and errors.

And then in 1978, a moon was discovered orbiting Pluto and

was given the name Charon. By seeing how quickly Pluto's gravity pulled Charon around in its orbit, you could finally measure its
mass directly. The result: Pluto was *not* ten Earth masses large. In
fact, it was nearly a thousand times *smaller* than Earth—ten times
smaller even than Earth's moon. (At that point, one wag plotted the
putative mass of Pluto as a function of the year the estimate was
made, and he published a joke paper predicting that, at that rate,
Pluto's mass would become imaginary in 1984!)

Tombaugh had discovered a body where Lowell had predicted
a planet should be. But Tombaugh's discovery was not confirmation of Lowell's predictions. Instead, it was just a nasty coincidence.
Pluto was a small body that merely happened to lie where Lowell
had wanted to find a much bigger planet.

Pluto's orbit was already known to be very different from any
other planet's: highly eccentric and greatly inclined or tilted compared to the plane of the other planets' orbits. But still, it was the
only known object out beyond Neptune. And Tombaugh himself
had looked for years, unsuccessfully, for any other body as bright
as Pluto out there. So even though we now knew it was a very
small body, there was no immediate need to redefine Pluto's status.
Schoolchildren continued to count (and identify with) the little eccentric planet at the edge of the solar system.

PAUL: Pluto is so far away; given how small it turns out to be, I can
see why it was hard for anyone to notice it or anything else out
there.

GUY: Its visual brightness is "fifteenth magnitude" (the bigger the
magnitude value, the fainter the object). That's ten thousand times
fainter than the faintest star you can see with the naked eye. Tombaugh would expose photographic plates for an hour at the telescope
to gather enough light for Pluto to show itself. A longer exposure
on any one spot would show fainter objects there, of course, but
then, the more time you spend on one small part of the sky each

night, the more nights it takes to search the rest of the sky. (And for technical reasons—it's called "reciprocity failure"—after a certain point you can't just assume that ever-fainter guys will show up with ever-longer exposures on photographic plates such as he was using; they are not so efficient at recording really faint objects.) In any event, with the equipment he had, there simply weren't enough hours of darkness in a night to expose a photographic plate that could record anything more than a few magnitudes fainter than Pluto. Tombaugh himself kept looking, for the rest of his life, and never found another object out there.

By the 1990s, however, electronic CCD chips had been invented that were far more sensitive than Tombaugh's photographic plates and, thus, capable of seeing far fainter objects. Searching for the source bodies of a particular class of comets, David Jewitt and Jane Luu discovered the first of what are now called the Trans-Neptunian Objects, or TNOs. These bodies, presumably balls of ice and rock a few tens or hundreds of kilometers across, had been predicted, by a number of scientists (most famously Gerard Kuiper) many years earlier, to be the source of a well-known family of comets. Over the next ten years a thousand such "Kuiper Belt" objects were found.

Was Pluto just another TNO? Not really. Though all of them followed eccentric, inclined orbits in the same region of space as Pluto, the TNOs were clearly much smaller even than Pluto. Where Pluto's diameter was around 2,300 km, the TNOs were only a few hundred kilometers or less across. Their brightness was roughly twentieth to twenty-fifth magnitude, hundreds to thousands of times fainter than Pluto. That's why Tombaugh hadn't seen any of them in his searches.

Still, some scientists saw the writing on the wall and began to agitate for the IAU to redefine Pluto's status. It was clear that, for practical purposes, keeping Pluto's data with the TNO data made more sense than keeping it with the planetary data. However, in 1999, the executive council of the IAU insisted that no change in Pluto's status was being contemplated.

But soon even that stance began to change. Special observing programs had been set up to look systematically for asteroids or comets likely to hit Earth, and in the process they also found slower-moving asteroids in the outer solar system. In the year 2000, the Spacewatch project found a TNO with a diameter of over 1,000 km. That was comparable in size to Pluto's moon Charon and about as big as the largest known asteroid, Ceres. Still, it was treated as another asteroid (albeit in the Kuiper Belt, not the asteroid belt), given an asteroid number (the round number 20000 in honor of its special size), and the name Varuna, for the Hindu deity often associated with the Roman god Neptune.

Two years later, Michael Brown and Chad Trujillo at Cal Tech found an even bigger TNO, about 1,200 km in diameter. This was clearly bigger than Ceres (the largest of the traditional asteroids in the belt between Mars and Jupiter) but still only a fraction of Pluto's size. It was given another nice round number (50000 . . . lots of smaller asteroids had been found in the meantime, thanks to projects like Spacewatch) and the name Quaoar, the creator-god in the traditions of the Tongva tribe (native to the area of California near Cal Tech).

Then, in 2003, a remarkable object, eventually named Sedna (and numbered 90377), was found at twice Pluto's distance from the sun, with an orbit that would take it out to nearly a thousand times Earth's distance from the sun. In order to be seen, Sedna's size must be approaching 2,000 km, twice the diameter of Ceres. It began to become clear that Pluto's standing as the largest object beyond Neptune was in danger of being overtaken.

Sedna had also been found by Mike Brown and his team at Cal Tech. They proposed the name after the Inuit goddess of the sea and suggested that other similar objects should also be named for characters in the mythologies of people who lived in the arctic region. By then, it was clear that many more such objects would be discovered . . . Sedna didn't get a special number. But if more objects were out there, odds were that some of them were likely going to be Pluto-size or larger. And so the IAU decided to take a second look at the issue. They formed a committee, and I found myself on it.

WHITEOUT

PAUL: Is it just me, or is the light getting odd out here? It's been starting to cloud up, and now for some reason I'm beginning to feel a bit dizzy. I don't know why . . .

GUY: It's a whiteout beginning. Since the ice and the clouds are exactly the same color, you can't tell where the snow ends and the sky begins. The horizon disappears. So you lose all visual clues about where you are. Without a sharp boundary like that, you can start to get disoriented.

When the clouds cover up the sun, and the illumination becomes diffuse, the ground is lit equally in all directions. You don't see shadows on the ground anymore, and without the contrast of shadows, the texture of the white snow underfoot becomes harder and harder to make out. Soon you have a hard time even knowing where to put your feet, where to avoid ridges and hollows in the snow. It's actually rather dangerous to try to walk around in conditions like these. We'd better get back into the tent.

PAUL: Ah, that's better. With the little camping stove lit, we can start to get warm and maybe have some tea or hot chocolate. It's great to be out of the cold!

Speaking of cold, back to Pluto. So, you were saying, your committee were the guys who demoted Pluto?

GUY: Actually . . . no. The committee I was on was only one of several such committees put together by the IAU. That right there predicts a disaster in the making.

What's more, the original membership expanded and expanded again until, with nineteen people (both ex-officio officers of the IAU and other scientists heavily involved with the study of Pluto and similar bodies), we actually never were able to meet all together at the same place and time. We broke into ad hoc subgroups of members who happened to be together at a given international conference, who would come up with some shaky compromise definition of what a planet was, only to be ignored by the next subset who happened to be together at the next international conference.

It was a hopeless task. Along with the intractability of the problem, personal animosities and egos worked against coming up with a general compromise.

PAUL: Speak for yourself!

GUY: I am. Certainly I know my ego played a role in my own arguments. I was tempted to push my position over somebody else's in part just for the ephemeral honor of being "the guy who invented the definition of a planet." And I think I could see the same temptation playing out in other members of the committee as well.

The best our committee could come up with was an arbitrary 1,000-km radius boundary; but no one was very happy with that rule, and it did not have anything like overwhelming support. Worse, there were three different ways of working out the details for how to apply this boundary, and no one of them got a majority vote.

Our committee heard dozens of suggestions for defining what a planet was, from folks on the committee and from outside. Everyone had an opinion. In fact, I would say that of all the discussions since then, I haven't heard any definition proposed that we hadn't already thought about within the committee. Likewise, we also heard cogent reasons for why every proposed suggestion was fatally flawed.

Some proposals suggested that a body would only be a planet if it had an atmosphere, or active surface geology. But making those observations for faint, distant objects is impossible to do at present; at best these objects are mere dots of light in our telescopes. Do we have to wait for a hundred years' improvement in telescopes and spacecraft before we can know how to classify a newly discovered body?

There is clearly a sharp divide in size between Pluto and the next smallest planet, Mercury, so some division based on size might work. But what size? And who's to say that every system of planets discovered around other stars will have such a convenient cutoff point? And what happens if a body bigger than Pluto but smaller than Mercury is discovered out beyond Sedna?

Should we insist on a definition easy enough to explain to a schoolchild—or at least one that won't make us the laughingstock of late-night talk shows?

What you called a "planet" depended on what kind of work you did. If you study the physics of what goes on inside a planet, Earth's moon is a perfectly valid example of a planet. If your study is about how planets behave in their orbits and how they're perturbed by other bodies, even Mercury is considered somewhat questionable as a planet.

But here's the important point: the reason we couldn't give the IAU Executive Committee a definitive statement of what "everybody thinks" a planet should be defined as, was because *there was absolutely no consensus within the astronomical community itself.*

PAUL: I think I can see one of the problems here. The way you classify things—the criterion or rule that you employ to say whether or not something belongs in a group—depends on what it is you are trying to accomplish. You could sort a library of books alphabetically by author or title, if you wanted to make it easy to find a particular book or a particular writer. Or you could sort them by subject matter, if that's how you needed to search among them.

GUY: I know someone who shelves his books by the color of their covers, since he usually can't remember the title or the author of the book he's looking for, just what the cover looks like.

PAUL: Some say that our goal, in setting up categories, should be to "cut nature at its joints"—our category divisions should correspond to real category divisions that exist in nature. But others say that we have no way to know nature's real category divisions: when we set up categories, it is purely a matter of what is convenient and useful for us. And still others say that there are no real category divisions in nature.

There is the famous claim (somewhat controversial) that Eskimo languages have many more words for snow than do other languages. The idea was that people who live and work constantly with snow will be more attentive to subtle differences in types of snow, will have a greater variety of uses for snow, and will, therefore, use subtler and more nuanced categories to distinguish various kinds of snow. I'll admit that as I look around me here in Antarctica, I'm just seeing snow—I'm not seeing all that many different types of snow. But maybe during your long stay here hunting meteorites you developed a keener sensitivity to various kinds of snow.

GUY: Certainly when you are riding a snowmobile over sastrugi, your rear end develops a certain sensitivity! (Sastrugi are ridges of snow turned hard when the wind erodes them—and I do mean hard. The word is German from Russian dialect, not Eskimo.)

PAUL: Some people argue that all category divisions are artificial, that there aren't sharp lines. For example, they say, if you add sand to an anthill grain by grain, it's hard to identify the moment when that one extra grain of sand turns it from an anthill into a mountain.

GUY: Sometimes that's called "fuzzy logic."

PAUL: Does that mean that there is no real or clear borderline or distinction between an anthill and a mountain? Or is it like the horizon outside here in Antarctica during a whiteout, where there is a real borderline or distinction, but we can't see it, we don't know where it is? Or does it mean that the borderline or distinction depends on us and on what it is that we want to accomplish?

GUY: The anthill example is relevant to the growth of asteroids, which are piles of rubble. I don't know about anthills or mountains, but what happens with asteroids is different from what that argument suggests.

First, the physics. We know that most asteroids today are loosely accumulated piles of rubble. They are made out of individual rocks, and some of those rocks wind up arriving on Earth as meteorites; we can pick them up off the ground here in Antarctica, where they're easy to find. But the density of the individual meteorites—that's what we measure in our lab at the Vatican—is about twice the density of the asteroids they come from, which we can identify in our telescopes. We conclude that the asteroids are loose piles of these stones with a lot of empty space between the rocks.

Let's assume that an asteroid is growing by accumulating dust or pebbles or boulders; we're talking now about what was going on back in the early solar system, when there was lots of that stuff floating around the sun.

There are two different forces competing within such a rubble pile.

On the one hand, each pile of rubble is held together by its own gravity. The more rubble, the greater the force of gravity. As it happens, the strength of the asteroid's gravity also increases as its density increases; a compact, not very "rubbly" pile will have stronger gravity than if the same amount of stuff was accumulated into a loose, fluffy pile of rubble.

On the other hand, the individual rocks all have a certain internal strength. That strength, and the friction between the rocks, tends to hold the bits of rubble apart, against the force of gravity.

In a rubble pile, the strength of the rock is greater than the local strength of gravity; each rock maintains its own shape and keeps the spaces between the rocks open. But as the rubble pile gets bigger, the gravity gets stronger, while the strength of the individual rocks stays the same. So you can see that eventually there will come a size, a critical threshold, where, finally, the local gravity is strong enough to break the rocks down and squeeze out the spaces between them—at least at the center of the pile, where the pressure is strongest.

But notice what happens here. Once the squeezing starts, the density of the rubble pile grows, and so the local gravity grows, which means that the squeezing goes even faster, making the density increase even faster, increasing the gravity even more, and the whole shebang collapses, all at once, into a smaller but denser new form. There is *not* a general gradation from more fluffy to less fluffy. If the rocks are strong enough to resist gravity, they resist. The moment they aren't strong enough, then the whole system collapses. It could very well be that one grain of sand on the pile, or some other relatively small change, is enough to trigger the cascade.

PAUL: Well, this suggests an obvious way to define a "planet," then. Before the collapse, it's an asteroid; after the collapse, it's a planet.

GUY: Yup, you'd think. This kind of definition was, in fact, one of the definitions proposed to the IAU. It has the even bigger advantage that it defines not only two distinct states of existence, but it also gives you an obvious reason you would want to choose this particular dividing line. Piles of rubble don't have interesting geology going on inside them. But once the body is big enough to be solid, it can behave cohesively like a planet, with tectonics and maybe internal melting—behaviors that you can see on their surfaces in the forms of faults and flows—the kinds of structures that structural geologists know all about studying. So knowing if an object is a rubble pile or a coherent body tells you which "tools" to pack in the toolbox you bring to study the object.

This kind of definition has another advantage. A solid body like this, controlled by its self-gravity, will have a smooth, round shape. Rubble piles are more likely to look like moldy potatoes. So you'd expect that you could spot from a distance which ones are the solid objects and which ones are rubble.

PAUL: Problem solved, it seems to me.

GUY: Not quite. For one thing, the collapse point depends on what the body is made of or how hot it is. If you had two identical-looking bodies, but one was made of relatively soft ice and the other of hard iron, you might ask why the ice body should be a planet while its iron twin is not.

And surely a body can't be a planet only when it is warm (and able to flow into a smooth, spheroidal shape) but then stop being a planet when it cools down and becomes rigid. Sounds far-fetched? In fact, that precise example describes the case of 4 Vesta, the second-largest body in the asteroid belt. In its early history it was molten and round. After it froze up, repeated large impacts turned it into a very odd, misshapen lump with a large dimple at its south pole. So is it "planet" or not?

But there's a different issue, as well. If you want to know whether an object is a planet because you want to study its geology, this definition is fine. But other people besides you are studying planets, for reasons completely different from yours, and to them this definition is utterly irrelevant. The people who study how planets are actually formed, or how the entire system of planets evolves over time, look at the issue with completely different criteria.

To some, the issue is how the body was formed. Did it come from a substellar lump of gas and dust, or from an accretion of smaller rocks? We can't actually see how Pluto or its neighbors formed, but we can make some well-educated guesses. At the very least it seems clear that Pluto did not form in the same manner that Neptune formed; why else would they be such utterly different sizes?

Others worry about how stable a planet's orbit is. Clearly, if we are talking about two different kinds of bodies here, it makes no sense to lump them together into one classification. If for no other reason, we have to keep track of a small body's orbit more carefully than a large body's, because it is more prone to be moved about by the bigger bodies. But there's an even more subtle point here. When Neptune does change its orbit (ever so slightly, due to perturbations from Saturn or Jupiter), it perturbs Pluto, as well; it drags Pluto along. But if Pluto gets perturbed by some other body, it doesn't stay perturbed; Neptune drags it back.

So, should planets be only those bodies whose orbits are hard to perturb? But then it might be that a given body would be counted as a planet if it is close to its star, where its orbit is more rigidly fixed by the star's gravity, while the same body would not be a planet if it were farther away from its star and, thus, more prone to having its orbit altered.

PAUL: So this is not so much the problem of the anthill versus the mountain. The problem is not that there aren't any clear criteria to employ in distinguishing planets from non-planets. The problem is that there are several different clear criteria that could be employed. You'd like for them all to agree with one another, but they don't. And you get murky results.

To me it is sounding more and more as though the distinction between "planet" and "non-planet" runs along the lines of the distinction between "healthy" and "unhealthy." Deciding whether a particular person is "healthy" depends on a multitude of factors. You might have healthy blood chemistry but poor eyesight and hearing. Does that make you healthy or not? It depends.

It seems that we're looking for a simple answer (planet or non-planet) to a question that is too complex for a simple answer.

GROUND BLIZZARD

PAUL: Speaking of complex . . . what is that racket we're hearing? Over the past few minutes it's gotten noisier and noisier outside this tent!

GUY: The wind has picked up. It's starting to blow snow across the ice, and that's what's hitting the side of the tent. It might go away in a bit, or it might keep this up for days. You can't work outside when the weather is like this; we were pretty much stuck inside our tents for twelve days running when I was down here in 1996.

The wind always comes from the South Pole—it's called a katabatic wind. You can think of it as the cold air from the pole sliding down the slope of Antarctica until it reaches the sea.

PAUL: Noisy, cold air always coming from the South Pole . . . Hmm, is that anything like noisy, hot air flowing out of the committee room when you scientists get into arguments about the status of Pluto?

GUY: Very funny. And while all those arguments were going on, the planet discoverers hadn't stopped discovering new worlds. By the summer of 2005, it was clear that a certain TNO first found in late 2003 had been studied enough to establish that it was sufficiently far away, and yet sufficiently bright, that it certainly had to be bigger than Pluto. (It was designated, for the time being, 2003 UB313; the code indicates the year, half-month, and order in which it was discovered.) And two other newfound bodies, only slightly smaller, were waiting in the wings.

These bodies created a crisis for the IAU. It needed to make a decision. But its original commission had failed to decide.

Knowing that there was now an intense public interest in "defining a planet," the Executive Committee came up with a new commission, this one much smaller (five members plus a chair) made up of scientists, science journalists, and historians. Their deliberations were held in secret, to prevent the kinds of jockeying that had caused the first commission to fail. The decision of that commission was finally announced, as a press release, at the beginning of the two-week IAU General Assembly (GA) in Prague in August 2006.

The first stage in the process at the GA, however, occurred even before the proposed definition was published. At the opening session, the IAU officers proposed an alteration in the way that proposals would be voted on. Instead of being voted on by representatives from each member nation, they proposed—and the GA agreed—that "scientific matters" (i.e., not budgets and other issues relating to the business of the IAU itself) would be subject to a vote of the individual IAU scientist members themselves. In this way, they hoped to make sure that any controversial matter, such as the definition of a planet, would have the fairest possible hearing and, whatever outcome resulted, the broadest possible support.

The small committee's proposal for defining a planet was announced on the Wednesday of the first week of the GA. It defined a planet in just the way I described above: it is a body whose gravity was strong enough to overcome its own individual, internal forces and produce a spherical shape, approaching what the geophysicists call "hydrostatic equilibrium." By that definition, Pluto was not the only small body to earn the title *planet*; the largest asteroid, Ceres, and the newly discovered big Kuiper Belt objects would all be planets.

It also proposed that if a planet's moon was big enough that the pair's center of gravity lay outside either body, both should be called planets in a binary pair. This is the case of Pluto and Charon. Thus, along with the then-yet-to-be-named 2003 UB313, the number of planets was to be expanded to twelve, with, undoubtedly, more to come.

But as we have seen, while this definition described the intrinsic properties that a geologist might look for, it said nothing about the orbital dynamics of a planet. That didn't make the celestial-dynamics group happy at all. A group of them prepared an alternative definition, which they presented on Friday of the first week at an open forum on the topic. They kept the geologists' definition but added the idea that a planet also had to "dominate its region of space" gravitationally. In other words, it had to be big enough to cause a noticeable perturbation on the star it orbited and on the orbits of the other big planets. In addition, they argued that a true planet would have to be a body that had "cleared out" its region of the solar system when it was formed.

There were about three hundred planetary scientists present at the session where this definition was proposed, and the opinion among them was overwhelmingly in favor of adding this dynamical constraint. (Along with being the president of Commission 16, Moons and Planets, that year I also served as the recording secretary of the entire Planetary Sciences Division, and so I took the minutes of that session.) Even the people attending that session who didn't like the new proposal had to admit that some 60 to 80 percent of those present approved it.

But these dynamical considerations were hard to explain, especially in terms that would make sense to people who were not dynamicists—much less the general public or the pundits who would be making fun of us on late-night TV talk shows.

PAUL: I am surprised and happy to learn that you were worried about the need for things to make sense to the general public. I had been assuming that the only considerations that came into play were those internal to astronomy and the needs of astronomers.

But what you are saying is that, in coming up with a new definition of a planet, astronomers tried to take into account not only their own needs but also the needs and limitations of the general public.

I think that shows a good, realistic, mature appreciation of the fact that science cannot and should not operate in isolation from the rest of society. After all, much of the funding for science comes from the public, so you need to keep the public "on board" with what science is doing. And often findings and discoveries in science end up affecting how society functions. And since future astronomers will come from the public, it can't hurt to present astronomy in enticing terms that they can understand.

But I'm wondering if the quality of science will be compromised if it caters too much to the needs and limitations of the public. If you "dumb down" the level of your terms and categories to the point that the public can understand them, will that end up inhibiting the conversation among scientists—the internal conversation that you yourself say is so important? It's hard to have a good conversation if you have to keep using "baby words."

And I don't think Isaac Newton worried much about adjusting his discourse to the level of public understanding!

GUY: It was up to the IAU Resolutions Committee to come up with the final wording for the definition of a planet. They were a small group of astronomers, none of them planetary scientists, charged with preparing all IAU proposals to be voted upon. And that's the second place where the Vatican connection appears, because the chair of that committee was none other than our Vatican representative, Father Corbally.

Father Corbally and the IAU Resolutions Committee tried their best to simplify the proposed language, so that the general public would get the idea. They published their version of the new proposed definition on the following Tuesday, the second week of the IAU General Assembly. But the dynamicists felt that their ideas had been oversimplified. That Tuesday at noon, the president of the IAU, Ron Ekers, hosted an hour-long general-forum session on the topic. The session turned acrimonious; indeed, there were dark intimations that the IAU Executive Committee was somehow trying to push a definition on to the assembled astronomers that was contrary to what the astronomers themselves believed.

That evening, a second open public session was held. But this time, instead of being chaired by the president of the IAU, it was hosted by the member of the Resolutions Committee who is arguably the most well-respected astronomer in the community: Dame Jocelyn Bell Burnell.

As a graduate student, Jocelyn Bell had discovered the first pulsar. There was an international outcry when her thesis advisor, Antony Hewish, not she, won the Nobel Prize for Physics for that discovery. But she was in the forefront of those defending the decision, pointing out that she had merely followed Hewish's directions; he had planned the experiment and interpreted the results, and so she insisted he, not she, really deserved the prize. In a delightful irony, the result of her gracious intervention ultimately meant that she became more famous than most prizewinners, for *not* winning the Nobel Prize!

At that session, Dr. Bell Burnell did a masterful job, listening and explaining. After hearing everyone out, the group prepared new versions of the proposal for the final General Assembly vote on Thursday afternoon.

I remember entering the hall for the Assembly that afternoon. Each of us was checked to see if we were wearing "Press" or "Guest" or "Member" badges; only if you had a "Member" badge were you given a yellow card, to be raised when a vote was taken. I would guess that easily a thousand people attended, more than half of them voting members.

First, four technical resolutions on other matters were voted on; they passed without controversy. Then the Pluto resolutions came up for a vote: Resolutions 5 and 6.

The text of the resolutions had been distributed beforehand, so everyone had a clear idea of what was coming up and what each vote meant, but to explain things in person onstage was Jocelyn Bell Burnell, along with Father Corbally, who had prepared the final text, and Richard Binzel, the chair of the "small committee" whose original proposal had been shot down.

Each resolution was split into two parts. Resolution 5a divided all the objects in the solar system into three classes: planets, dwarf planets, and small solar-system objects. This was discussed at some length and then taken to a vote. The vote was overwhelmingly in favor; no precise tally was needed. (I, too, voted in favor.) It was clear to everyone that objects like Pluto were definitely different from objects like Neptune but also different from the rubble-pile asteroids.

Resolution 5b was the kicker. It added the word *classical* to the first category of planets. To make clear exactly what this one word would mean, Bell Burnell brought out a large ball, representing the larger eight planets; a handful of small pebbles; and, between them, a box of cereal and a stuffed yellow plush dog toy, representing Ceres and Pluto (think about it . . . the goddess of grain and the famous Disney character!). Then she placed an umbrella over the ball, the cereal, and the stuffed dog—but not the pebbles. The intent of this change, she said, would be to include both the large eight bodies and the dwarf bodies together under the umbrella term planet. The large ones would be "classical" planets; the small ones, "dwarf" planets.

This was the crucial vote. Vote "yes," and Pluto (and the others) would be planets, while a "no" vote meant they were not. The "ayes" stood to be counted, with several hundred votes in favor. But when the "nays" stood, it was clear that they overwhelmed the "ayes." Everyone present agreed that no count was needed. (I voted "aye" and lost.)

Resolution 6a noted that Pluto was the first of a special class of dwarf planets located out beyond Neptune (as opposed to Ceres, which was also dwarf-planet-size but located in among the asteroid belt). This passed on a counted vote, 237 in favor, 157 against. (I voted in favor.) Resolution 6b explicitly named these bodies "plutonian." It narrowly failed: 183 votes in favor, 186 against. (I voted in favor, and my side lost.) Perhaps this resolution was considered more "technical" than Resolution 5; in any event, many members abstained.

I think one reason "plutonian" lost was the sense that the word itself was inelegant. "Plutonian" sounded too much like the element "plutonium." Other suggestions had the same problem: for example, "pluton" already has another meaning in geology, and it is already the French name for Pluto!

PAUL: For people who aren't scientists, it could seem mighty strange that these matters were settled by votes. They might wonder: Wait a minute, I thought science was supposed to be objective. Since when is scientific truth settled through votes and politics?

But that's not what I'm taking away from what you said. It wasn't the size, shape, and qualities of various solar-system bodies that were up for a vote. Everyone could agree on that. What was up for a vote was how best to categorize them.

And in this case the goal of categorization was quite pragmatic: you want to establish categories in the way that will best support future work in science. What that says to me is that when working scientists make use of a set of classifications and categories, they aren't claiming to "cut nature at its joints"; they aren't assuming that the categories and divisions that are employed now in their field correspond to real categories and divisions in nature (though they hope that they do). Nor are they assuming that the same categories and divisions will continue to be used in the future. The categories and divisions that are employed now in a field are the working conceptual tools of that field.

The hope is that the tools that are being used are the ones that will best foster future progress. And the hope is that the tools won't have to be changed too much in the future, because change is difficult. That makes sense to me, and it clears things up for me.

Though the category "planet" has great historical importance, as far as working scientists are concerned, that category is a tool. In deciding to change the definition of "planet"—in deciding to change the nature of the tool—astronomers had to think about what tool would be best and most apt for their work. But they also had to take into account that the term "planet" has a long history and is well known by the public. The bottom line is that the term "planet" denotes a conceptual tool, not an honorific title. And, therefore, there was no intention to "demote" Pluto. (But somehow I still feel bad for poor Pluto!) So, what happened after all the votes?

GUY: Those present at the IAU General Assembly, including those (like me) whose "side" had lost, felt that we had, indeed, had a fair hearing. You had to be there to understand the passions on this issue among IAU members . . . and to be impressed with how well they did, given those passions.

But following all the press attention (it was August, and the "silly season"—the slow-news season when newspapers are desperate to fill their pages), a bunch of American planetary scientists who hadn't attended the General Assembly vote—indeed, many of whom had never participated in the work of the IAU—decided they didn't like the fact that their pet planet had been "demoted" (that was their term), and they started circulating a "petition" asking that the matter be "reopened." Left deliberately vague was any proposed improvement to the definition. It reminded me of my experience on the first, stalemated committee: you can easily get a majority to agree that "something better" is possible, but you'll never get a majority to agree on what that "something better" is!

And, practically, what possible change can such a "petition" effect? It gives the impression that somehow this vote represented an important scientific decision, that science itself is decided by votes, and that the outcome of science can be influenced by political agitation. But as you've already noted, that's not the case. This vote was not an important scientific decision; the point of the vote was not to decide whether something was scientifically true or false. Only good data can make that decision for you. The point of the vote was to adjust the categories that we use in such a way as to give working astronomers the tools they need to communicate well with one another and to make good discoveries in the future—and to do so taking into account the needs and limitations of the public.

To me, the final definition makes scientific sense. The division between asteroids and dwarf planets was very much needed. My own research into meteorite and asteroid densities shows a distinct difference between small but compact objects like Pluto and the loose rubble piles of asteroids. But the division between dwarf planets and big planets also makes sense, from the point of view of the dynamicists. And one advantage of this definition is its creative ambiguity. In reply to the question, "Is Pluto a planet?" it will be equally true to say, "Yes, it's a dwarf planet" and, "No, it's a dwarf planet." That reflects the ambiguity of nature itself.

I am satisfied that the procedure was as fair as the IAU could have made it, and that the final definition is one that will adequately allow the IAU's Division III (or, as it has since been renamed, Division F: Planetary Systems and Bioastronomy) to get on with its work. That, after all, is what the vote was all about. It's an IAU decision for IAU's purposes. The IAU knows (as it said explicitly at the General Assembly) that it can't control the definition of words in all the world's languages, or how the general public uses those words. It can only define terms so as to determine how it goes about distributing its workload to appropriate committees and working groups.

And, finally, 2003 UB313 was given a name. Following this vote, the IAU accepted the proposal by its discoverer, Mike Brown, that it be named for Eris: the goddess of discord. Eris now has a number (136199) and a place in the catalogue of solar-system bodies, along with a hundred thousand other asteroids and TNOs . . . including Pluto, now, as well.

THE HUMAN TOUCH

PAUL: Inside a tent like this, sipping a cup of hot chocolate while the wind blows a blizzard outside—what a great place to tell stories! I guess you had lots of time for storytelling while you were down here.

The story you've been telling about "demoting" Pluto both builds up and tears down my confidence in science. It tears down my confidence in science because it shows the degree to which doing science is like making sausage. It involves politics, personalities, egos, multiple agendas, and all sorts of human considerations that are far removed from the purity of an individual scientist making observations and analyzing data.

But it builds up my confidence in science because it shows that no one person gets to have the final say-so. Modern science is a cooperative, competitive venture, with lots of people involved, with lots of checks and balances. The process can be messy and political. But in the end a scientific result gets accepted as true only after it has been put to the test, very thoroughly, by lots of different people and groups.

And you know what? The dirty historical secret is that many religious doctrines came to be via similarly messy, political processes. The "data" that the Church has to go on are various personal encounters with God, as recorded in Scripture, as passed down in Church tradition, and as reflected on by Popes and by ecumenical councils. At times the Church has chosen to "codify" something about those encounters in the form of doctrine and dogma.

Usually the Church has done that only when troubling controversy was brewing: you don't go to the fuss and bother of coming down decisively on the question of whether or not Pluto is a planet unless ambiguity on that question has been causing conflict and confusion. And you don't go to the fuss and bother coming down dogmatically on the question of what it means to say Jesus Christ was the Son of God unless ambiguity on that question has been causing division and confusion—as was the case by the time of the First Council of Nicaea. The Nicene Creed, which was produced by that council, wasn't intended as a summary of every important Christian belief. It focused on issues that were divisive and controversial. You'll find no mention in the Creed that God is Love. That's a very important Christian belief. But since it wasn't controversial or divisive, the Church didn't bother making a dogmatic proclamation on that issue.

GUY: Notice something else. From the time it was proposed that there was another planet outside Neptune, to when Pluto was discovered, to when Pluto was finally measured and found to be just a little ice ball, our understanding of Pluto and the words we used to describe it have changed a lot. Just remember that description in the Heinlein science-fiction novel from 1958 I quoted. In some cases, the change occurred because we learned more about the object; in other cases, the change is that the words we used a hundred years ago, such as "planet," mean something different now than what they meant when people first used those words.

The same thing happens, of course, in religion. How many kids today have heard the phrase "Holy Ghost" instead of "Holy Spirit"? Our way of describing God constantly changes and grows, both as we understand God better and as the words themselves take on new meanings.

But while all those things have changed, the truth itself does not change. Pluto itself has not changed. God is the same as He ever was. If there was any change, it was in us.

PAUL: And as we saw in both the case of Pluto and the Council of Nicaea, the fight over finding the right words to describe that truth can be ferocious.

Here's what I find interesting. The loudest fighting in the Pluto story was about how to define categories: planets, dwarf planets, etc. But it sounds as though there was much less fighting about giving names to individual objects. I would have expected louder arguments about names than about categories: names usually remain forever, but categories come and go. Once a name has been given, no one will try to change it. (Can you imagine the uproar, among both scientists and the general public, if there were a serious proposal to rename Pluto?)

And it seems that only some astronomical objects are given names, while most others are merely given identifying numbers. What's up with that—is there politics involved? For example, why is it that Charon, Varuna, Quaoar, Sedna, and Eris were given names instead of only numbers? Was there much politics involved in their naming?

GUY: Well, as it happens, I am also on the IAU committee that names features of planets. We have rules and guidelines, which takes much of the controversy out of the decision. But, yes, controversies still come up.

It's one thing to say that Nobel Prize winners in physics deserve to have craters on the moon named for them. But what do you do if one such physicist was also a fervent and outspoken Nazi?

Then there's the tension between having names that honor people we all agree deserve the honor, and having names that are unambiguous. We want to avoid more cases like the situation that already exists in which there's a crater named Copernicus on the moon and another crater with the same name on Mars. But should Mercury—where crater names honor artists—name two different craters Manet and Monet? I always get those painters confused!

Finally, we want to be sure that we represent and honor all the different cultures and nations on Earth, not just those who happen to have representatives on our committee. But even best intentions can go awry; names that appear innocent to our eyes can have religious or political significance in other cultures.

As to which features on a planet get names, the answer is, again, pretty utilitarian. We only name features when they need names, for instance, if someone is writing a scientific paper about a particular region and needs a name to refer to a feature that they're talking about in their work.

PAUL: Now that I think of it, isn't there an asteroid somewhere out there named "Consolmagno"? How did that happen?

GUY: Asteroids are named by a different committee, not mine!

All asteroids start with a discovery number: the year when they were found, followed by letters that indicate when during the year, and then more letters and numbers in the order of when it was reported to the IAU's Minor Planet Center. But, of course, it turns out that something found this month may actually be an object reported many years earlier, for which we didn't yet have enough good observations to fix its orbit. Once the orbit of a given object is fixed, however, it is given another, permanent, number. That's its official designation.

However, asteroids can also have names. The folks who discover asteroids have the right to choose any name they wish, with rather loose rules. Pop stars and fictional heroes—and living people—are all acceptable. Military or commercial entities or pets are not. That was one reason given (tongue-in-cheek!) for why Pluto could not be an asteroid: Pluto is Mickey Mouse's pet dog, after all!

Since there are hundreds of thousands of known asteroids, it's not too hard to find one that can be named to honor somebody. As it happens, every three years the asteroid, comet, and meteor community has a general scientific meeting, and at the banquet, a number of our community are honored by having asteroids named for them. That's how I got my asteroid, which is officially known now by both its number and its name: 4597 Consolmagno. The fellow who discovered it is a colleague and friend. It's a small community.

PAUL: I'm still fascinated that you astronomers wanted to take the public's needs and limitations into account as you were redefining the term "planet," and there are a couple of things I want to note.

First, there's an interesting irony going on here. Back in the 1600s, during the controversy involving Galileo and the Catholic Church, a question on the table was how to interpret some passages in the Bible that seem to say that the Earth is motionless at the center of the universe. At the time, no one wanted to say that the Bible was wrong—everyone wanted to be able to say that the Bible was correct in all regards. But science seemed to show that the Earth was in motion and not at the center. Galileo argued that the biblical authors had *accommodated* the relevant passages to our limited human understanding.

That is, the biblical authors knew that they were speaking to common people, and accordingly they spoke to them in terms that they could understand and that accorded with their daily experience. And so they spoke to them in terms of a non-moving Earth. *Accommodation* became an important principle of interpretation of the Bible. It is used to help figure out when to interpret the Bible literally and when to interpret it figuratively.

The astronomers of the IAU, in coming up with a new definition of "planet," were trying to come up with a definition that would be more-or-less *accommodated* to the understanding of regular people. They were willing to use a definition that would be somewhat less precise or useful, from the perspective of experts in astronomy, if that would help keep the general public informed and interested about astronomy.

But here's the ironic result. When future historians of science try to interpret the categories and definitions that were used by the astronomers of the early twenty-first century, they will have to keep in mind that those categories and definitions were (in some measure) *accommodated* to the understanding of the common people of our time! In the same way, biblical scholars have to keep in mind, when they interpret the Bible, that its language was (in some measure) *accommodated* to the understanding of the common people of those times.

Once something has been accommodated to the public at large, the public at large will feel some ownership for it.

The idea that Pluto is a planet doesn't belong only to astronomers. It belongs to the culture at large. The planets have a place in the heritage of our culture: you can find references to Pluto and the planets in nonscientific stories, books, poems, and songs. You can see Pluto and the planets portrayed in paintings, in historical displays in planetariums, and even in stained-glass windows.

When it was decided that Pluto was no longer a planet, after all, that had the effect of rendering a network of cultural references anachronistic, from all those mnemonics we employed to teach our kids the names of the planets, to even the Mickey Mouse character, Pluto, who got his name in 1931 soon after the planet was announced. That comes at a cost. That's part of the reason that there's still some resistance in nonscientific quarters to the IAU's recategorization of Pluto as a non-planet. And that's part of the reason there was resistance in nonscientific quarters to Galileo's effective recategorization of Earth as a planet.

GUY: Meanwhile . . . have you noticed, the wind has died down? I think it may be safe to go back outside. C'mon, pull on your boots and big red coat, and follow me back outdoors.

There's another reason we've come here to Antarctica for our chat about Pluto—a reason that has to do with the nature of science and, indeed, with the nature of "truth as we know it."

I mentioned before that visiting Antarctica is about as close as you can get to being on another planet. So just stand here for a moment, and take in what you see. Feel it in your bones. The clouds rushing off to the horizon, the sky turning blue again. The white hills and the vast expanse of blue ice, gently rippled and crested with snow, looking like frozen breakers at the seashore. The remoteness. The emptiness.

Breathe deeply, and tell me . . . what do you smell?

PAUL: I don't smell anything.

GUY: Exactly.

Where have you ever experienced air this clean, this pure? No diesel fumes, no pollution. No flowers or trees or cut grass. No salt, no fish, nothing rotting, nothing growing.

Nothing.

Standing in Antarctica is like standing in the midst of a plain of pure data, the ideal of a scientist. It's breathtaking.

But it's also frozen. And dead.

Even when you're out here in a tent and haven't bathed for six weeks, you still don't smell much; the cold keeps the smells down, I guess. But when you get back to the big main base at McMurdo, where a thousand people live (at least during the summers; it's down to a few hundred during the winter), the atmosphere is quite different. In the prefab dorms you have a constant, musky tang of wet woolen blankets and the presence of too many people crowded together into too little space. But it's alive.

Science is not data. Science is people living with, living in among, the data. And, yeah, we bring our own distinct human tang to everything we touch. It's not pure. But it's what brings it alive.

PAUL: Your story of how Pluto got redefined has given me a strong sense of the tang of human in science. And how much I miss that tang of the human presence, standing here.

Well, the controversy about Pluto showed how much people cared about it. So I guess I don't feel so bad for poor old Pluto, after all.

GUY: You know, when it was just another planet, Pluto was always an ugly duckling. It was the wrong size and in the wrong kind of orbit. Every time I would teach a class giving an overview of the solar system, I would have to always add, "except Pluto . . ." All planets are in nearly circular orbits, except Pluto. All planets are in orbits that lie close to the same plane, the plane of the sun's equator, except Pluto. All planets keep to their own region of space and don't cross the paths of other planets, except Pluto.

But now that Pluto has been defined as "dwarf"—a fascinating and important category of solar-system objects that until recently we never even knew existed—it's no longer an ugly duckling. It's one of a whole family, among dozens of similar objects. And, indeed, it has become the new standard against which all the other dwarf planets are defined.

It is now a beautiful swan.

Day 3: What Really Happened to Galileo?

SETTING: THE TOWER OF THE WINDS, VATICAN CITY

MAPPING THE WORLD

PAUL: Today we're back home in the heart of the Vatican, in a part the tourists usually don't get to see, the tower above the Vatican Secret Archives. I wanted to get us someplace warmer than yesterday, so I decided that getting back to Italy would be ideal. At first I thought of going to Florence, where Galileo lived most of his life, but, instead, I decided this would be the perfect spot for our chat about Galileo: the Meridian Room in the Tower of the Winds.

GUY: In this room in 1582, a Dominican priest named Ignazio Danti set up a mechanical weather vane, which was quite a clever device for the sixteenth century: a vane on the roof of the tower was connected via rods and gears to an arrow on the ceiling of this room, which would point the way the wind was blowing. The walls were painted with frescoes depicting the winds in biblical settings: a ferocious storm on the north wall, for example, and on the south wall a stiff breeze blowing across the Sea of Galilee, where Jesus walks on the waters.

However, our theme today is not Galilee, but Galileo.

PAUL: By the way, that's not the first time someone has made a pun on Galilee and Galileo. Another Dominican, a friar named Tommaso Caccini, did that in a sermon given in 1614 at Santa Maria Novella in Florence. Caccini's sermon was probably the first public attack by a churchman on Galileo for supporting the idea that the Earth travels around the sun: the first move in the infamous "Galileo Affair." Caccini quoted from Acts 1:11, where an angel, speaking to Jesus's disciples, says: "Men of Galilee, why do you stand looking up toward heaven?" Caccini made this into an attack on Galileo and others for putting trust in scientific observations of the heavens instead of in what is said in the Bible.

A lot of people wonder, wasn't there anything Galileo could have said or done to avoid his problems with the Church? Why didn't the Church tolerate or understand what he was trying to say? Why did this whole conflict occur? Where did things go wrong?

GUY: By coming to the Tower of the Winds, are you making a snarky comment about Galileo's lack of skill in judging which way the winds were blowing from the Vatican?

PAUL: Nothing that subtle, and nothing that crass! I want to start here mostly because this room in some ways represents a good moment in the relationship between the Church and science. Here we had, in Father Danti, a priest who was also a scientist—interested in the winds—and a techie who was good at mechanical gizmos. He was also an artist, a very skilled painter; he was the same guy responsible for the famous map rooms you walk through in the Vatican Museum, just a few floors below where we're standing.

GUY: Oh, yeah, that long room full of different maps of Italy, wonderful mash-ups of geography, politics, and art. I always love walking through those rooms. On one of the maps he even marks the little village that my great-grandfather came from.

PAUL: The most famous part of this room in the Tower of the Winds is its floor: notice the meridian line, a line scored in marble running due north and south, marked with the classical astrological symbols for the different constellations of the zodiac.

Now, if you look up there on the south wall, you'll see a little *putto* up in the sky with his lips pursed, blowing the wind across the Sea of Galilee. But his mouth is actually a small hole in the wall, lined up precisely with this marble line on the floor. Sunlight gets focused through that hole like light through a pinhole camera, and an image of the sun is created on the floor of the room.

Precisely at solar noon, the image of the sun crosses the meridian line. As the seasons change and the sun passes higher or lower overhead, the position of the image on this line moves north or south. The astrological symbols along the line indicate the astronomers' best determination of what group of zodiac stars the sun must be sitting in, when its image shines on that particular part of the floor.

GUY: *Camera* is the Italian word for "room." So this room in the Tower of the Winds is, literally, a sixteenth-century solar camera. Of course, the zodiac position of the sun is calculated, not observed, since you can't actually see a constellation when the sun is in it.

PAUL: By determining where the sun hits the meridian line on the shortest and longest days of the year, which will be the northernmost and southernmost ends of the meridian line, you can calculate the middle of those two extremes. That tells you where the sun ought to be on the equinoxes, the first day of spring and fall.

And the vertical angle between that equinox spot and the hole in the wall provides a precise measure of your latitude. A number of churches around Europe have meridian lines; the famous astronomer Cassini used them to make the first accurate map of France, for instance. You can find a history of meridian lines in the book *The Sun in the Church: Cathedrals as Solar Observatories,* written by J. L. Heilbron, a history professor at the University of California, Berkeley.

GUY: So, you said this tower was built in 1582. Wasn't that the very year that Pope Gregory XIII issued his famous reform of the calendar? Did they use the observations on this very meridian line to calculate how much they had to shift the calendar to make the seasons line up properly?

PAUL: Actually, no. By the time this tower was built, the calendar commission had already worked out how to fix the calendar. This meridian line wasn't used for any of the calculations. But it was used to demonstrate to important visitors how such calculations were made, and to show how much the days of the calendar had to be shifted.

The calendar reform, like this beautiful tower, represents a wonderful moment of concord between faith and science. The Pope hired astronomers, the best he could find, for the very practical purpose of fixing the calendar and making sure it stayed correct into the future. Part of the reason for doing this was to come up with a way to determine the date of Easter and other religious feasts that move with Easter every year.

Among the work the astronomers relied on was that of Copernicus, who had the best table of lunar and solar positions available at that time. This all took place years before the Galileo Affair. Among the people who served on that calendar commission was a Jesuit mathematician named Christopher Clavius, who wrote books explaining to scholars and the rest of the world how the new calendar system worked. Clavius later became one of the great reformers of the teaching of mathematics, writing books that everyone used for the next hundred years. He was also a good friend of Galileo.

INSANELY GREAT

GUY: We forget sometimes that Galileo was part of a large and very able scientific community, many if not most of whom were clergymen like Father Clavius. Galileo's achievements are remarkable, but they didn't occur in a vacuum.

It's important to recognize the importance of that kind of collaborative community, because we can see that at work even today. Everyone knows the names of people like Steve Jobs or Bill Gates or Mark Zuckerman who made it big in the world of high-tech, but none of those guys could have done a thing without a whole crowd of programmers and electrical engineers who came before them or worked with them. Where would Jobs have been without "Woz"? They were the ones who made the bold ideas actually work.

Consider, for example, the story of one such modern-day tech hero.

Leonard Gallow was born in Chicago in 1964, at the tail of the Baby Boom following World War II. His dad taught music at Roosevelt University, which is a nice little school in the south part of the Loop; though it doesn't have a big national reputation, it's known in theater circles for its music and drama departments. And young Leo had musical talent, too.

As is typical with musical kids, Leo also had a great facility with numbers. He later recalled that, while one of his earliest memories was watching the moon landing, by contrast the musical gathering at Woodstock (which happened less than a month later) meant nothing to him. (Of course, love and peace and rock and roll don't mean so much to you when you're just five years old.)

He went to the science-math high school in Chicago, Lane Tech, graduating in the class of 1982, and got a scholarship to the University of Chicago. But the University of Chicago was not really a good fit for him. He was bored with all the required humanities and philosophy classes. Though he started out as premed, soon he discovered that his real passion was computers. He wound up spending a lot of time with his friends from high school who were attending the University of Illinois Circle Campus; some of them worked in the Electronic Visualization Lab there and got him connected up with Project PLATO and "p-notes," an early version of e-mail.

After his junior year, he dropped out of the U of C to make money writing software, freelance, for a series of small companies in the Chicago area. In order to keep his library and other privileges at the University of Chicago, he took a part-time position as a math and computer-science tutor there. (Along with the required philosophy courses he hated, the college also required calculus courses, hated by the philosophy majors.) He was tutoring a class on linear algebra that morning in 1986 when the *Challenger* shuttle blew up. Meanwhile, he wrote a lot of the software for the nascent PC industry that went into financial-planning programs like Quicken.

His software was good, but Chicago wasn't the place to make it big. He had dreams of fame and glory. He had thought when he left school that his software would quickly make him rich and famous, like Bill Gates, but that hadn't happened. As his thirtieth birthday began to loom before him, he was consumed by the fear that life was slipping past him. So in 1992, when he was offered a chance to move to Boston and work for a start-up on Route 128, he jumped at the chance.

The fact that the work was mostly writing code for military projects didn't faze him; he was learning on state-of-the-art computers from some of the best cryptographers in the world. (He'd gotten the job because of his connections at Chicago with Adi Shamir, a pioneer of modern cryptography and cryptanalysis—the S in RSA, one of the first public-key encryption algorithms.) Besides, it paid well. But it still wasn't going to make him famous. He was from the Midwest, and like many midwesterners he had an insatiable hunger for "Making It Big." Eventually he quit the company (taking a good chunk of stock with him) and set off on a series of his own projects.

For the most part, his bright ideas did not pan out. He was a good programmer and great at self-promotion but not necessarily smart with the business end of things. Then all that changed in the spring of 2009.

During the downturn of the Iraq war, he was having lunch at the Echo Bridge with the legendary engineer Tom Crowder and a couple of his buddies from his military-industrial days, talking about their projects. Of course, it was all classified, but he'd worked with them and had the same clearances. It was never clear how much of Leo's great idea came from things they said or just things they implied that he was able to develop on his own; what's certain was that, by the end of that conversation, he knew he'd finally come up with his "Great Breakthrough."

The financial collapse of late 2008 helped. He had never kept money in the market, but once stock prices were at fire-sale levels, he bought big, and during the rebound over the next few years he had enough to live on without having to keep a day job. It took more than a year to work out the bugs, but by 2009 he had a working model in hand.

Well, you know what happened next. There's hardly anyone around who hasn't been touched by Gallow's Gizmo (as the blogs like to call it). He originally called it the Pocket Lawyer; its first slogan suggested, "Keep a good lawyer in your pocket!" The size of a cell phone, it could scan text or use voice recognition, and, no matter how complicated the jargon, it would perform a complete syntax analysis (using the kind of recognition algorithms devised by the NSA). It could tell you, in plain English, what a contract or a lease agreement actually meant—where the loopholes were and who had to pay what.

But more revolutionary was the claim that his Gizmo could even tell you if the person you were Gallowing was telling the truth. (Or, at least, if they thought they were telling the truth.)

It wasn't just software; Gallowing required using Leo's own special hardware, his Gizmo, which only he professed to understand how to make . . . though soon there were certainly a lot of imitators trying to follow his lead.

When the Gizmo was rushed to market in early 2010, it caused a sensation. Within a year Leo had enough fame and money to write his own ticket. Mayor Daley invited him back to Chicago to set up his Gizmoplex in his hometown, and Leo jumped at the chance. (Even while living in Boston, Leo had kept ties with family and friends in Chicago; he had tutored the mayor's kids, and he used to shoot hoops with a young Barack Obama at the University of Chicago.) Soon he was a regular on *The Daily Show*, his Twitter feed had more than four million followers, and he had to hire a small staff to maintain his Facebook pages and blog.

Of course, his Gizmo didn't come without controversy. It threatened to put a lot of lawyers out of business, and obviously they're the last people you want to make enemies of. They know how to make problems for you.

But, worse, there were certain aspects to his Gizmo that made a lot of religious fundamentalists queasy . . . once people started running things like the King James Version of the Bible through his analysis. It appeared to suggest that large chunks of many favorite Bible passages were meant to be taken as fiction, and it gave legal interpretations of certain rules that diverged from long-accepted teachings. This made Leo particularly unwelcome among Bible Belt folk in the Deep South. And as sales of the Gizmo spread beyond the shores of the United States, including versions in other languages, it became one more point of conflict in a world already racked with instability. Having a Gizmo that ran Arabic, the language of the Koran, became illegal in much of the Middle East. Meanwhile, the fact that President Putin of Russia actually endorsed it, and claimed the Russians had invented it first, didn't make Leo any more popular at home.

So, how did this all play out? What happened to Leo Gallow and his great Gizmo?

I'm writing this in 2014, so I can't say. Leo's just turning fifty, entering a comfortable middle age, and doing quite well at the moment. Will Leo's entanglements with the political powers of 2014 do him any good in 2023, as he approaches age sixty? What about ten years later, in 2033, when he'll no longer be the young guy shaking things up but an old man with a history, for better or worse? How will the "culture wars" in America play out, and how will that affect his popularity or the value of his stock options? Will he wind up being ensnared in wars and revolutions in the Middle East—or elsewhere? We don't know yet.

All right, I'll come clean: I am making this up. Leo Gallow is a fictional character. But I've built into Leo's story a series of point-by-point parallels with the life of the person we are really interested in here. Galileo was born exactly four hundred years earlier than "Leo," in 1564; was educated in the 1580s; was a math teacher in the 1590s; and worked on developing military technology until he had his big breakthrough in 1609, when he used a telescope to look at the heavens and was the first to see craters and mountains on the moon, the phases of Venus, the stars that make up the Milky Way, and the moons of Jupiter. His book describing these discoveries, *The Starry Messenger,* came out in 1610 and was an immediate sensation. Galileo and his telescope became famous all over Europe. He leveraged his fame to gain better jobs and financial success—and he also made enemies. The peak of his fame probably came with the publication of his book *The Assayer* in 1623; the low point came with his trial in 1633, when he was almost seventy years old.

Why did I include a made-up fairy tale about Leo Gallow in our discussion of the very real Galileo? To make a point. Galileo's life was also long and complicated. All sorts of things had gone on in his world before he ever got around to making his telescope. He had opinions about politics and religion. He had favorite music and favorite food. He had a girlfriend—and three kids—whom he left behind when he moved back to Florence. I think it helps to put his life into the same sort of context as our lives . . . to ask at each stage of the Galileo story what were (or will be) our hopes and concerns when we were (or will be) Galileo's age?

When it comes to Galileo, many people think they know what

happened—they think they know the story already. But they don't know. So the reason why I blended fact and fiction in my story of Leo Gallow was that I wanted you to experience that feeling of being a little confused, a little uncertain about what was true and what was made up. Get used to feeling that way, when dealing with the story of Galileo.

It's hard to separate truth from fiction in the Galileo Affair. Don't trust anyone's version of the Galileo Affair. Not mine, not what you thought you knew before you heard mine. And, yeah, I know it's irritating to have to face up to that, but there are some real puzzles to this story that we'll never stop arguing about.

This much I do know: if you want to understand the actual person, Galileo Galilei, and not just the mythical icon commonly referred to as "Galileo," you have to consider him in the context of his times. Galileo's long and complicated life extended across a history in which Popes and princes, geniuses and generals, were not merely names in history books but people he knew and interacted with.

JUST THE FACTS, MA'AM

PAUL: Wow, that was quite a wind-up! So then, what really happened to Galileo?

GUY: If you go to Amazon.com and search "Galileo," you'll find more than ten thousand results: *ten thousand* books that discuss Galileo! "Galileo biographies" give you hits for nearly four hundred books. No, I haven't read them all. But I bet every one of them tells a slightly different story of "What Really Happened."

PAUL: And that's fine. History is more than a recitation of the facts of a case; it's selecting which are the important facts and trying to make sense of them. The "facts" are anything but simple!

And, of course, the answers you find depend on the questions you're asking. Various historians have asked different questions, depending on their own various interests, biases, or prejudices about science and about religion.

GUY: That said, it would help to start with at least an accurate record of the basic, undisputed facts. And that's something an awful lot of what's been written about Galileo fails to do. For example . . .

Here are some quotations about Galileo from a university-level introductory astronomy text that was widely used while I was teaching: "Italy was no place to introduce unorthodox ideas that might challenge Church teachings . . . in 1616 Cardinal Bellarmine interviewed [Galileo] and ordered him to end his astronomical work. Books relevant to Copernicanism were banned, including *De Revolutionibus* . . . [At his trial in 1632, Galileo] must have thought often of Giordano Bruno, tried, condemned, and burnt at the stake in Rome in 1600. Bruno had been an outspoken critic of the Church in many respects, but one of his offenses was Copernicanism . . . Galileo was sentenced to life imprisonment . . . he was held in confinement at his villa where he could meet his family, though other visitors were forbidden."

Much of this is just plain wrong. For example, yes, Bruno wrote about the Copernican system, and, yes, he was burned at the stake, but there is no evidence that he was burned for being a Copernican. (He did plenty of things at the time thought deserving of death by burning . . . such as denying the divinity of Christ. The Bruno case, like that of Galileo, was more complex than that simple recitation suggests.) Furthermore, Galileo knew darn well that, as the hand-picked court philosopher of Cosimo de' Medici, the duke of Florence, he was too well protected politically for anything like that to happen to him, even at the height of his trial. In fact, when you read what people were actually saying at the time of the trial, nobody seems to bring up Bruno at all. It seems highly unlikely to me that Galileo, or anyone else, would have seen a parallel between his case and that of Bruno.

Nor did Cardinal Bellarmine order Galileo to end his astronomical work—Galileo published a number of books on astronomy during the fifteen years after 1616. Nor was Galileo denied visitors besides family during his last years in his villa; in fact, among his visitors was a young John Milton (he later mentioned Galileo in *Paradise Lost*). Nor was *De Revolutionibus*—in English, *On the Revolutions of the Heavenly Spheres*—banned. It remained in circulation, but the faithful at Rome were instructed to alter their copies to indicate that the heliocentric system was merely a useful mathematical fiction, not a description of reality. (And as Owen Gingerich describes in *The Book Nobody Read*, virtually nobody outside of Italy bothered to make those changes. In fact, even within Italy, where the decree supposedly applied, half the copies of that book remained uncorrected. At the Vatican Observatory, we have a copy dating from 1617 that we got from the Vatican Library itself, and it is not "corrected.")

PAUL: That's not to say that the people representing the Church who prosecuted Galileo were without blame. But accounts like what you've quoted from that astronomy textbook misrepresent the nature of their mistakes. And that's lamentable. If you don't identify the problem correctly, you are more likely to make the same mistake again, all the while thinking yourself safe from the supposed fault that you were never likely to commit in the first place.

GUY: If you want to get the undisputed facts in the case, my favorite source is *The Galileo Affair,* a book by the historian Maurice Finocchiaro, published by the University of California Press. What makes this book the place to start? Finocchiaro doesn't try to give an overarching explanation of what happened or why, though he does that in other books. Instead, he presents the important documents of the case—what Galileo wrote, what his friends and his enemies wrote, including the transcript of his trial in 1633—translated into English. You can read them for yourself and play amateur historian. And Finocchiaro's book also provides a helpful time line of all the major events of the Galileo controversy.

At the end of this chapter I've included, in an appendix, a time line of my own. It expands on Finocchiaro's time line. It includes the major events in the Galileo controversy, along with some explanatory notes. And it includes some other events that were happening at the same time, especially events from the Thirty Years' War. As you'll see below, I think that those additional events play an important role in explaining what happened to Galileo.

It's a long, complicated story—that is the first thing you should notice, if you look at the appendix. You'll see just how long it takes for the whole Galileo Affair to unfold. And you'll notice, as well, the various personalities involved, especially the family names of the different characters who keep cropping up (Barberini, Medici, etc.); family politics and rivalries play a part. And finally, you'll note the different locations where the story takes place. Geography is politics these days, and politics plays a central role in this story. So take a look at the appendix now, if you like. Or refer to it as needed along the way.

THE SCIENCE OF HISTORY

PAUL: OK, I looked at your appendix. Yeesh, did you know all of that off the top of your head? Amazing.

But, yes, sure enough, you've listed some of the important undisputed facts about the case of Galileo and his historical context. You said a lot. But you could have said more, and you could have said less—you made a selection of the data.

Anyway, out of all that information, where do you want us to focus?

GUY: The time line shows that the simple picture you get in a lot of textbooks gives a completely wrong impression. Galileo was not some outsider who got stomped on by the Church for having radical ideas. He was an insider. He had an important and visible position in the Medici court. He was friend to princes, cardinals, Jesuits, and the Pope. For more than twenty years, starting in 1610, he made a good living off his connections and his ideas, selling popular books (written in Italian, not Latin) and performing regularly at the Medici court and at salons throughout Italy—the "talk-show circuit" of his day.

The time line raises some interesting questions: It was in 1543 that Copernicus published his book, with his theory that the Earth moves around the sun. Why did it take seventy years before anyone thought the Copernican system was theologically dangerous? Why did Galileo snub other astronomers who considered themselves to be on his side, like Johannes Kepler (see 1609 in the time-line) and Horatio Grassi (see 1623)? How important was the role of particular individuals in keeping Galileo safe—people like Prince Cesi (see 1624 and following) and Grand Duke Cosimo II (see 1608 and following)?

But the most interesting question raised by the time line, at least for me, is this: Why is it only in 1632–33 that the roof falls in on Galileo? He had skirted trouble easily for the previous twenty years. But in 1632, the same Pope who was his fellow countryman and friend in the 1620s, and whose personal censor approved his book in 1631, not only refuses to defend him but plays a central role in personally seeing to it that he is tried and found guilty of heresy (or, at least, "vehement suspicion of heresy"), even rejecting Galileo's plea bargain. So what happened in 1632 that changed the Pope's mind? Why was he so adamant not even to accept a compromise or plea bargain?

And then, after all that, why was Galileo suddenly allowed to return home and write more books?

PAUL: OK, you've built up to this—you've posed the questions. What's your take? What do you think?

GUY: There are lots of answers. Too many different answers. That's why there are so many books and articles written on the topic. Which is to say, nobody knows for sure. On the face of it, what happened to Galileo doesn't quite make sense. Everybody who's ever looked at the story in detail seems to have come up with a different explanation. My suspicion is . . .

PAUL: (Rolling his eyes, tapping his foot) Ahem.

GUY: What's the problem?

PAUL: You've compiled a chronological list of facts that you take to be relevant. And now you're going to come up with some theory that pulls it all together. But the facts you have chosen as the dataset to be explained might or might not all be truly relevant. Your selection of those facts is colored by the explanation you want to give. So who is to say you're not just another astronomer doing the same thing as that astronomer who wrote the textbook you just slammed?

No, no, go ahead. I want to see where this is all leading.

GUY: Thanks. I think.

OK, here's my take on the subject. I admit, I am no more likely to have the full story than anyone else. I'm not a professional historian. But I have seen, close up, both how scientists can behave and how people in the Church can behave. That gives me a certain vantage point, at least.

I think we can agree that something happened in 1632 that turned Pope Urban VIII against Galileo. What was going on in Rome then? Or in Europe, for that matter?

The Thirty Years' War.

In those days, most of central Europe—present-day Germany, Austria, Czech Republic, Hungary, and all the other countries around there—was divided into dozens of principalities loosely affiliated under a Holy Roman emperor, who was elected by those princes. But with the Reformation, those principalities were split into Catholic and Protestant camps. In 1618, a meeting between representatives of powerful forces among both Protestants and Catholics in Prague turned ugly, and two Catholic representatives were thrown out of a third-story window. They survived. Historians date the start of the Thirty Years' War to this "Defenestration of Prague." Two years later, in 1620, a pro-Protestant army was defeated at White Mountain, outside Prague, and a long series of battles continued throughout Bohemia, eventually spreading through Germany.

It would take a series of books to do justice to the politics of this war. It has been called history's first "world war," since it involved alliances of many nations and battlegrounds as far-flung as Brazil (where the Dutch briefly captured Pernambuco from the Portuguese). It transformed the face of Europe.

On one hand, it could be read as a war between Protestant and Catholic factions in Germany, fighting to control whose religion would prevail. But it was also a war of Spain against all comers.

Spain was the richest and most powerful nation in Europe at that time. It had great wealth from its colonies in the New World, and it had close family connections to many of the other ruling forces in Europe, most notably the Holy Roman emperor.

Those nations in Europe that feared Spanish power fought against Spain, either actively or passively. Spain claimed to be fighting on behalf of Catholicism. But Catholic France was among the biggest backers of the "Protestant" side. Spain controlled Naples and Sicily and, thus, had large armies on the Italian peninsula. Although the various Italian city-states were very leery of Spanish power, they were also reluctant to oppose Spain directly.

Take a look at the time line in the appendix, and notice how many crucial events in the Thirty Years' War were happening in 1632.

PAUL: OK, I can see that Galileo's crisis in 1632 happened just when Pope Urban VIII was under particularly heavy pressure from Spain. As you note in the appendix, supporters of Spain in Rome were accusing the Pope of not supporting their cause, the Catholic cause. They wanted him to send troops in support of Spain, and they wanted him to threaten the French with excommunication if the French continued to oppose Spain. They even accused him of being secretly a Protestant! But it would have been very difficult for the Pope to act against France, since he had close ties with France and owed his election as Pope in large measure to French support. But is this anything more than a coincidence of dates? What does it have to do with Galileo?

GUY: Well, one thought is that Pope Urban went after Galileo simply as a way to keep his opponents at bay, to placate them by going after someone who had irritated them for a long time. Maybe he was just tossing Galileo to the wolves.

But I think the real connection is Florence: the home of the powerful and rich Medici family, who were Galileo's sponsors and protectors. Remember all the connections between the Medici and Galileo? Ferdinand I appointed Galileo to his first teaching job at Pisa (see 1588); Galileo tutored the young Cosimo in mathematics (see 1605–1608); and Cosimo, after he became grand duke, appointed Galileo as his own court mathematician and philosopher (see 1610). And remember that the Medici family had strong ties to both Spain and France: Ferdinand I was married to Christina of Lorraine, the granddaughter of a queen of France (see 1588 and 1621); and Cosimo II was married to Maria Maddalena, whose brothers were the king of Spain and the Holy Roman emperor (see 1621).

So what are the Medici doing during the Thirty Years' War, when the French are financing or fighting on one side, the Holy Roman emperor on the other? Which side is their money backing? The answer, interestingly, is . . . neither—for a very good, very deliberate reason.

Just when the Thirty Years' War is reaching a critical phase, the Pope's ally in Tuscany is under shaky leadership. The Medici Grand Duke, Ferdinand II, is barely twenty years old; his Spanish-ally mother is dead; and his French-ally grandmother remains the power behind the Medici throne but is elderly and waning (see 1631). So, here's my point: by coming down hard on Galileo, the pet philosopher of the Medici who had been Cosimo's friend and tutor, Pope Urban VIII was able to send a strong political signal to both sides: France and Spain.

PAUL: What was the signal? Spell it out.

GUY: By squashing Galileo, Pope Urban VIII could assure the Spanish that he had leverage over the Medici and that he would keep their money out of the war. And doing it this way allowed him to give that assurance without actually committing his own resources to either side.

PAUL: So for you, the main cause of the Galileo Affair was the politics of the time.

GUY: Exactly.

I mean, think about this: institutions don't change their behaviors any more than individual people do; after all, institutions are made up of individual people. If somebody's good at something, he stays good at it (or even gets better); if someone has a propensity to make a certain kind of mistake, you will probably see him making it over and over.

In spite of what popular opinion would have you believe, the Church is actually pretty good at allowing theological and philosophical debate and even tolerating dissenting points of view. There are exceptions, of course, but you find that those exceptions are fiercely contested precisely because they are seen as being outside the norm of what is usually tolerated. Certainly Galileo's opinions could not have been any secret to anyone in Rome for all those years; he got away with his stance for more than twenty years, up until 1632.

On the other hand, there are plenty of mistakes that the Church does tend to make, over and over. You know as well as I do the typical laundry list that people always complain about . . . but certainly, one very common problem that crops up throughout history is the temptation for Church leaders to align themselves and make deals (often with the best of intentions) with local political powers. You can see this throughout history, from the Church's involvement with Constantine, emperor of Rome, in the fourth century (like the Galileo story, that history is a lot more complicated than most popular treatments will admit), to the "Christian Democratic" political parties in nineteenth-century Europe . . . and beyond. This was especially true when the leaders of the Church themselves came from prominent political families.

It seems to me that if you want to see where the Church went wrong with Galileo, you should look to how the Church typically has tripped up over the years. More often than not, that happens when it has been tied up in politics.

PAUL: So, to sum up: As you see it, there are two things that need to be explained. First, how could someone as famous and well-connected as Galileo ever get into such deep trouble? He should have been pretty much bulletproof and untouchable. Something extraordinary must have been going on, for a person of his fame and stature to be forced to make a humiliating abjuration and to be placed under lifelong house arrest. Second, why did the whole Galileo Affair blow up in 1632 to 1633, rather than at some other time? Galileo was not doing or saying anything in the years leading up to 1632–33 all that different from what he had been doing and saying publicly for more than twenty years. So there must have been something else going on in 1632–33 that caused Galileo's problems to boil over at just that particular time.

And as you see, the way to explain it is via politics. In 1632–33 Pope Urban VIII was caught in a web of political pressure and intrigue among Spain, France, and Florence. Galileo was very prominent, and he was associated with Florence—he was the Medici family's famous pet philosopher and mathematician, but he was not an explicitly political figure. By publicly humiliating Galileo, Urban was able to appease France and Spain by putting some needed distance between himself and Florence—and he was able to do this without making any sort of real political break with Florence.

DAMAGE CONTROL?

PAUL: What's interesting is that your way of explaining the whole Galileo Affair doesn't make any reference at all to science or to the relationship between faith and science—which is what most people think the whole thing was all about.

I'm not going to disagree that political considerations play a part in Galileo's downfall. But I'm going to offer a different explanation. I think that science and faith *were* involved—though not in the way that most people think.

But even if you and I offer different (perhaps complementary) explanations of what happened to Galileo, at least we agree that an explanation is needed! That has not always been the case.

I mean, look up at the ceiling of this room we're standing in, the Tower of the Winds. On the big compass rose for Father Danti's wind vane, each direction has a name besides just east or west; each wind was given its own name, the twelve winds known to the Greeks and Latins, and the eight names used by the Italians. Even today the Italians use these names; for example, they refer to the sirocco wind from the desert southeast, or the tramontana wind from over the mountains to the north. But in Father Danti's time, naming the winds was enough; they didn't ask what actually caused the winds to blow.

But you and I are children of the Enlightenment and children of modern science: we always want an explanation.

So here's my take. What happened to Galileo was not a confrontation between the (oppressive and villainous) Catholic Church and (free and heroic) modern science. After all, in Galileo's time, modern science as we know it didn't yet exist. Galileo straddled the gap between old and new ways of understanding science, as well as a new understanding of how science relates to religious faith.

As I see it, Pope Urban VIII was (or chose to be) deeply offended by a passage near the end of Galileo's *Dialogue Concerning the Two Chief World Systems*. Some years before, Urban had assured his friend Galileo that he could write about the two world systems, as long as he kept his discussion at a merely hypothetical level. This meant that Galileo was free to explore the relative merits of the two systems, to his heart's content. But he should avoid claiming to prove definitively that one system was better than the other.

Galileo's clever way of dealing with this limitation was to write his book in the form of a dialogue. Two of the characters in the dialogue, Salviati and Sagredo, repeatedly fall into discussions in which the Copernican system seems to come out as the clear winner over the Ptolemaic system. The third character, Simplicio, tends to go along with these discussions at first. But after a while, he usually admits that he doesn't really understand what the other two are talking about, and he reminds them to keep the discussion at a merely hypothetical level.

Simplicio's most explicit reminder along these lines comes near the end of the *Dialogue*. Salviati and Sagredo have been discussing Galileo's theory that the motion of Earth is the cause of the rise and fall of the ocean tides, with the oceans sloshing back and forth in response to that motion. Simplicio says:

> "I confess that your idea seems to me much more ingenious than any others I have heard, but I do not thereby regard it as true and conclusive. Indeed, I always keep before my mind's

eye a very firm doctrine, which I once learned from a man of great knowledge and eminence, and before which one must give pause. From it I know what you would answer if both of you are asked whether God with His infinite power and wisdom could give to the element of water the back and forth motion we see in it by some means other than by moving the containing basin; I say you will answer that He would have the power and the knowledge to do this in many ways, some of them even inconceivable by our intellect. Thus, I immediately conclude that in view of this it would be excessively bold if someone should want to limit and compel divine power and wisdom to a particular fancy of his." (*Dialogue Concerning the Two Chief World Systems*, translated by Stillman Drake.)

The "man of great knowledge and eminence" was Pope Urban VIII.

What Simplicio says here is just what Pope Urban had said to Galileo years before: we can't claim to know for sure which is the true cause of a particular physical phenomenon, because any phenomenon will be consistent with a variety of possible causes, some of which we know and some of which we don't know. Salviati immediately agrees.

So why did Pope Urban take offense? Shouldn't he have been pleased that Galileo included his words in the *Dialogue*? Well, no. First, Galileo put Urban's words into the mouth of the character Simplicio, who throughout the *Dialogue* functions as something of a straight man.

GUY: "Simplicio" had an honored role in philosophical disputes as a name invoked to present the response of a man whose honesty and simplicity insures him from falling into the traps of clever reasoning.

PAUL: Yeah, but while the philosophers might have known that, the folks buying Galileo's book would see Simplicio as the butt of a book-long joke. The *Dialogue* was not a scholarly tome written in Latin. It was written in Italian and sold to the mass market.

Of the three characters, it is Simplicio who most consistently admits that he doesn't really understand what is being discussed. Salviati and Sagredo always treat Simplicio with affection and respect but also with a hint of condescension and forbearance. Certainly any Pope—not least someone like Urban VIII, who was proud of his sophistication—would have found that a particularly uncomplimentary comparison to make!

GUY: *The Oxford English Dictionary* says that the first time the word "urbane" was used to mean "sophisticated" was in 1623, the year Urban was elected Pope. That may be coincidence, but certainly that's how Urban saw himself.

PAUL: I think this was a case of Galileo trying to be too clever once again. And to make matters worse, after Simplicio's remark, Galileo has Salviati respond with an "approval" that sounded a bit condescending and ironic. Not smart on Galileo's part.

Second, to anyone reading the *Dialogue,* it was clear that, despite his protestations, Salviati really thought that he had definitive proofs that the Earth moves. So from Urban's perspective, his friend Galileo not only disobeyed his instructions but also subjected both him and his instructions to a kind of subtle, winking ridicule.

The Pope was not amused—he took it personally. And his (former) friend Galileo paid the price.

GUY: Still, this raises another issue worth asking. Why should Urban VIII have been concerned about astronomy at all? Why did he want to put limitations on what Galileo could or could not publish—didn't that amount to a kind of censorship?

PAUL: In one sense, you could say that Urban VIII was subjecting Galileo to the same kind of censorship that modern scientists accept all the time, when they try to publish in scientific journals. Modern scientific journals won't publish just any old article; papers have to be vetted and approved by referees. Referees don't insist that only the articles that they agree with can be published; there's plenty of room for the hypothetical in scientific publishing. But referees do their best to make sure that nothing that is clearly wrong gets published. And that is a kind of "censorship."

GUY: I have been furious at referees at times; I have also been thankful when they've found serious problems in my work and stopped me from making a fool of myself in print. But it's self-censorship. When we referee papers, we scientists are censoring ourselves.

PAUL: Yes, this kind of refereeing involves scientists judging other scientists. But Pope Urban VIII was a member of the Academy of the Lynx, just like Galileo, and he was also a trained theologian—which Galileo was not. Urban may have considered himself as much an expert as Galileo on what passed for physics in those days—not to mention theology and metaphysics, where Galileo had little formal training.

Anyway, Pope Urban VIII felt that he had been disobeyed and subtly ridiculed by Galileo. But even more, I think he felt that Galileo had betrayed and undermined him,

Here's my take: I think that the limitations Pope Urban VIII asked Galileo to observe in writing his *Dialogue* were intended as part of a quiet strategy of papal "damage control" relative to the Church Decree of 1616. Urban VIII was known to have held that the anti-Copernican Decree of 1616 was a mistake on the part of the Church (see, for example, 1624 in the appendix). The fact that Urban gave Galileo permission to discuss the Copernican system in print, even hypothetically, shows that he was not particularly interested in enforcing strictly the 1616 edict that it was "foolish, absurd, and heretical" to say that the sun was motionless, and "foolish, absurd, and erroneous" to say that the Earth moves.

However, Popes are reluctant to disagree directly with their predecessors. I think Urban's quiet strategy for undoing the damage that had been done by the Decree of 1616 was to allow his friend Galileo, a person of great renown and rhetorical skill, to discuss the Copernican system in print, hypothetically, with urbane humor and disarming wit. That would help create the congenial rhetorical space and good will needed for the Church to be able to "back down" gracefully from the mistake it had made in 1616. Urban's goal never would have been for the Church to come out in favor of Copernicanism. But I think that he wanted the Church to be able to back down from having declared Copernicanism foolish and absurd.

Perhaps if Galileo had come through with the kind of book Urban was looking for, maybe Urban would have been able to back him against his detractors, saying (with a nod and a wink), "What's your problem? Galileo's treatment is merely hypothetical. Are you saying that you find yourself attracted by this merely hypothetical treatment?"

But Galileo didn't come through in the way Urban had hoped— Galileo didn't "play ball." His treatment of the two systems was witty, urbane, and persuasive—as hoped. But it was not hypothetical. Once Galileo's book came out, Urban realized that everyone could see that Galileo thought that the Copernican system really was correct. And this undermined Urban's damage-control strategy and backed him into a corner. It put him in the position of having to defend the Decree of 1616, which he didn't really want to have to do. And, on top of all that, it subjected him to ridicule, by putting his words into the mouth of Simplicio. Urban felt angry and betrayed.

According to you, Guy, Galileo should have been untouchable because he was so famous and well-connected. On my reading, it was precisely because Galileo was so famous and well-connected that Pope Urban looked to him for subtle help with his damage-control operation. But the book that Galileo provided was not the book Urban was expecting, and there was no way for Urban to "plausibly deny" that his friend Galileo had gone beyond the merely hypothetical.

GUY: So why wouldn't Galileo "play ball" the way Urban VIII wanted him to? Galileo was well capable of providing the sort of persuasive but merely hypothetical treatment of Copernicanism that Urban wanted to see. Why didn't Galileo stay within the limitations set forth by his friend the Pope?

PAUL: It may well be that Galileo thought that he had done so. At several points in the *Dialogue* he affirms that his treatment should be understood as being merely hypothetical. He probably hoped and expected that those explicit declarations would suffice, despite the fact that the content of the book revealed his clear commitment to the truth of Copernicanism. But the explicit declarations did not suffice; they had the effect of allowing people to portray Galileo as a clever and ill-intentioned dissembler.

GUY: Which, in fact, he was.

PAUL: A clever dissembler? Yes. But ill-intentioned? I don't think so.

But there was something about Galileo's personality that just wouldn't allow him to make any argument merely halfway. Galileo had to go for the kill, every time. He did it in an elegant and witty fashion—but he always went for the kill.

Look how he went out of his way to offend several important Jesuits. The Jesuits originally had been his supporters, but after Galileo's interactions with Christoph Scheiner and Horatio Grassi, they were no longer in play as possible defenders when he got into trouble. When Galileo got into trouble, there was no shortage of people who were willing to pile on—people who were looking for a little bit of payback.

You've cited in the appendix Galileo's acerbic satirical style in *The Assayer,* but that's hardly the only example. Here's another, which is my favorite. In the *Dialogue,* Galileo provides three main arguments that the Earth moves and that the sun is motionless at the center: the argument from the observed retrograde motions of the planets, the argument from the observed changes in the path of motion of sunspots across the face of the sun, and the argument from the rise and fall of the ocean tides.

Let's look for a moment at the second argument, the one from sunspots. Observing the sun with his telescope, Galileo noticed that sunspots tended to move from left to right across the face of the sun, over the course of 12.5 days. And he noticed that the path of motion of sunspots across the face of the sun changed in shape in a regular way over the course of a year. In certain seasons of the year, sunspots moved across the face of the sun in a straight line, either from lower left to upper right or from upper left to lower right. But in other seasons of the year sunspots moved across the face of the sun in a curved path: sometimes curved downward, sometimes upward.

Galileo realized that the observed seasonal changes in the paths of sunspots across the face of the sun could be explained very simply, if . . . the Earth moves annually around the sun; the sun rotates on its own axis in a period of twenty-five days; and the sun's axis of rotation has a tilt of around seven degrees relative to the axis of the Earth's revolution around the sun.

Now if Galileo had simply proposed this possible explanation and left it at that, his opponents would have been reduced to sputtering silence, because there is no way to account for the observed motion of sunspots in a way that is consistent with Aristotelian physics. But, no, Galileo couldn't leave it at that. He had to go further. In his *Dialogue,* he essentially says to his Ptolemaic opponents:

"Hey, have you heard? Sunspots move across the face of the sun, and the shape of the path of motion changes over the course of the year. Interesting, huh?

"Oh, you didn't know about that? I was wondering, since I haven't heard anything from you as to how the Ptolemaic system could explain this newly observed phenomenon.

"Well, it so happens that I've already figured out for you what the best possible explanation would be for the motion of sunspots in the Earth-centered system. It turns out that you have to give four distinct motions to the sun: in addition to an annual motion around the earth and a rotation on its axis, there have to be two distinct 'wobbles' of the sun's rotational axis, one happening once each day and the other happening once each year. The problem, I'm sure you'll see, is that, according to Aristotle's physics, which you take to be true, it's impossible for a single body, like the sun, to have several different motions at the same time. So, it turns out that it's just not possible to explain sunspot motion in an Earth-centered system while remaining faithful to Aristotle's physics.

"Well, there it is. I'm glad I could be of help. And by the way, in case you're interested, the sun-centered Copernican theory can explain sunspot motions with no problem."

Galileo's style of argument humiliated people. He didn't simply let you know you were wrong. He took your argument, made it stronger and better than you ever could have made it yourself, and only *then* did he let you know it was wrong. He had to be the smartest guy in the room, and he had to let you know it, in ways that were clever and funny but often humiliating. (According to Mario Biagioli, in his book *Galileo, Courtier: The Practice of Science in the Culture of Absolutism,* this was part and parcel of the preening and self-promotion of court culture that Galileo lived in. But Galileo dealt it in spades.)

Anyway, when Galileo got into trouble, it was payback time.

GUY: Your theory sounds good, too. Maybe we're both right. Or maybe we're both completely off base. We could each write a book;

then, instead of 400 theories on Amazon.com, there could be 402 theories for what happened to Galileo.

If there's anything we can take away from all this, it's that the Galileo story is a lot more complicated than a lot of descriptions want to make it. It's not a simple case of finding proof that shows one side was right and the other was wrong. But that's true of everything in history. It's true in science, too.

THE HISTORY OF SCIENCE

PAUL: And that brings up one more important side to this story: it's about the change in how science itself worked back then.

It seems obvious, in retrospect, that Galileo was right about so many things. That can lead us to conclude that anyone who opposed him was either stupid or ill-willed—or both—and that the Church was fundamentally opposed to science. But that's not the case.

Yes, some of the people who opposed Galileo were probably stupid or nasty . . .

GUY: He could be stupid and nasty himself, at times. Like a lot of scientists I know!

PAUL: And, yes, there were tensions and misunderstandings between Church officials and the newly emergent scientific elite. But science itself was changing so rapidly at that time that people had trouble keeping up.

In Galileo's day, everyone knew that there was supposed to be an impenetrable disciplinary boundary between astronomy and physics. Physics (which at that time was called natural philosophy) dealt with the true nature and behavior of things in the real world; it was understood to be purely qualitative, and mathematics played no role in it. Astronomy, by contrast, was understood to be a branch of mathematics; it used geometry and arithmetic to calculate and predict the apparent positions and motions of objects in the heavens, as seen from Earth. Astronomy was supposed to deal only with "appearances"—with the apparent positions and motions of the stars and planets—and not with what they actually were.

GUY: And there had been a good reason for this separation. In Aristotle's time, those who followed Pythagoras thought you could describe nature in terms of mathematics, but to them that was more like "magic numbers" than careful measurements. Just as harmonic notes came from simple ratios for the lengths of harp strings, for example, they argued that planets had perfectly circular orbits with distances that could be expressed in a simple, regular, mathematical order—"the music of the spheres," they called it. Of course, it's not true. Aristotle, correctly, argued that imposing this sort of math on nature obscured the truth. But in the process, he obscured the places where math really could be useful.

PAUL: Galileo's life and work had the effect of challenging the barrier between the disciplines of physics and astronomy. He was appointed both philosopher *and* mathematician at the court of the Duke of Tuscany. That was strange enough. But his telescopic observations led him to discuss the real physical nature of the moon, and in his *Dialogue* he proposed two different physical proofs concerning the real motions of the planets.

To Galileo's contemporaries it would have seemed that he was running roughshod over what everyone knew should be the impenetrable boundary between the disciplines of physics and astronomy. When Bellarmine told Galileo in 1616 to stick with being hypothetical, he was simply asking him to stick with doing astronomy the way everyone else did it. It is not surprising that some of Galileo's contemporaries concluded that he was wrong, or that he was up to mischief.

GUY: And in some cases, he was wrong. And in some cases, he may have been up to mischief.

None of that can obscure the fact, though, that on the whole, he was right . . . in a way no one had encountered before.

PAUL: And another thing. In the Aristotelian thought world of Galileo's day, a scientific theory had to be "demonstrated" to be true: it had to be shown as being not merely *probably* but *certainly* true, the way that a mathematical theorem can be "demonstrated." That's the standard Galileo thought he had to meet in doing science . . . and it led him to overreach in the *Dialogue*, claiming that he had demonstrations when really what he had were good probable arguments.

In Galileo's time, no one had a demonstration that the Earth did *not* move. But since that was the default position, the burden of proof was on those who wanted to change it. In the minds of Galileo's contemporaries, like Cardinal Bellarmine, a demonstration would be needed to force a change in the default. Galileo had good *probable* arguments that the Earth moved. But he didn't have a demonstration. And in his day, demonstration was still the standard that was required in science.

In contrast to that, modern science accepts "high probability" as sufficient cause to accept a theory: Given the available empirical evidence and the various background theories that are already accepted, which of the proposed theoretical explanations accounts for the phenomena better than the others? That's the one that's good enough for us. And, for the same reason, we're happy to change our theories when one that's better comes along.

By the way, the word "probable" was also used differently in Galileo's day.

GUY: The mathematics of "probability" only began with the work of Pascal and Fermat in the 1650s, and statistics were another hundred years in the future. Instead, in Galileo's day an explanation was called "probable" even if it was less likely to be true than other explanations, as long as some reputable authority supported it.

PAUL: So that is where Bellarmine was coming from. It was assumed that demonstration was the standard of proof required for accepting an explanation in physics, and it was assumed (rightly!) that Galileo could not produce a demonstration of Copernicanism.

In Galileo's day, our modern notion of how to go about justifying physical theory on the basis of empirical evidence was not yet on the radar screen. We take for granted the importance of "crucial experiments." We take for granted that if you hold a particular theory, and a carefully conducted experiment shows decisively that the theory is wrong, you should abandon that theory. Galileo's new philosophy of science was a crucial step toward that modern belief.

Think of the famous story about how he dropped objects with different weights from the Tower of Pisa. Galileo was performing a crucial experiment, designed to refute the (then-taken-for-granted) theory that an object's speed of fall is proportional to its weight. But the whole notion of a crucial experiment works only if you understand nature as being rigidly consistent in its behavior—only if you think that the natural world acts like a kind of machine, with the behavior of objects conforming to a set of physics laws that are obeyed always and everywhere.

In Galileo's day, the idea of the world acting like a machine was very new. Most of Galileo's contemporaries took it for granted that nature should be understood as behaving not like a machine but rather like a living thing or organism. Organisms have regular patterns of behavior: acorns usually become oak trees, crickets usually chirp, and I usually get up at 6:00 a.m. That is how those things behave, usually and for the most part. But not always. Some acorns do not become oak trees, some crickets do not chirp, and sometimes I get up at 8:00 a.m.

If you understand nature not as being like a machine that acts always and everywhere the same way, but rather as being like an organism that acts the same way only usually and for the most part, then the whole idea of a "crucial experiment" makes no sense: you expect there to be exceptions to the rule, because nature has a built-in variability.

Finally, just as there was a strict disciplinary boundary between astronomy and physics in Galileo's day, so, too, there was a strict boundary between theology and philosophy. Galileo regarded himself as a philosopher and mathematician. He learned a lot of theology. But—crucially—he was not *licensed* to teach or discuss theology publicly.

In his *Letter to the Grand Duchess Christina,* Galileo tried to do some theology: he proposed methods and rules for biblical interpretation. His goal was to show that the motion of the Earth should not be seen as contradicting Scripture. The methods and rules Galileo proposed were good ones—Pope Leo XIII and Pope John Paul II acknowledged as much, centuries later. But those methods and rules were being proposed by a non-theologian—someone who was not licensed to teach or discuss theology publicly. Galileo sought to circumvent this limitation by distributing his *Letter to the Grand Duchess Christina* only privately, but it was circulated so widely that it might as well have been published. From the perspective of his contemporaries, Galileo was trespassing mischievously and illicitly on the turf of the theologians.

GUY: And in those days of the Reformation, with self-styled theologians inventing new religions right and left, that was not only impolitic, it was downright dangerous . . . not just for Galileo, but for everyone around him. New religions started wars—like the Thirty Years' War.

THE AFTERMATH

GUY: We can argue whether the trial of Galileo was primarily fueled by politics, or a failed damage-control operation, or by changes in the methodology of science, or by personal animosities. But the result remained the same: the heliocentric model for the universe was condemned in 1616 by a committee of cardinals, speaking for the Church. And they were wrong to do that. And that condemnation was the basis of Galileo's trial, nearly twenty years later. So that raises the question: when did the Catholic Church finally accept the heliocentric model?

Of course, there is no simple answer to that. It depends on what you mean by the Church, by "the model," and by "accept." There were always people, including important people in the Church, who accepted the idea of a sun-centered system. Nicholas of Cusa, a cardinal writing in the 1400s, talked about every star as a sun with planets. Obviously even in medieval times nobody found anything particularly heretical about the heliocentric system. Copernicus himself was a Catholic cleric who dedicated his 1542 book to the Pope, and it was not challenged until 1616, more than seventy years later, by enemies of Galileo.

Even after the Galileo trial, Catholics continued to be well familiar with the Copernican system. Jesuit educators throughout Europe—we have their textbooks and notes for their schools in Germany—taught the Copernican system in their schools. It was taught hypothetically in mathematics classrooms, as a mathematical technique to predict planetary positions, and not as a "cosmology" to explain how the universe worked. But it certainly kept knowledge of this system alive for a generation of scholars.

PAUL: And that's no surprise; as we have noted, before Galileo, astronomy was always considered a branch of mathematics, a calculating tool, not a description of reality. In Galileo's day, people working in the universities presupposed and understood rigid boundaries among the following disciplines: mathematical astronomy, which was used to calculate and predict the positions of the planets in the sky; natural philosophy, which dealt with physics and the real nature of things; and theology. But among the general public these boundaries may not have been well understood or may not have mattered. Galileo himself played on this ambiguity. He remained in contact with university professors and sought their approval, but he held a position not in academic university circles but at court, as philosopher and mathematician to the Grand Duke of Tuscany. And he wrote books intended not for academic specialists but for the educated general public.

GUY: We can see today how science popularizers such as Carl Sagan and Michio Kaku are treated by their professional peers, who cringe at the simplifications they employ when they educate the public in science. This antipathy is fueled both by a legitimate irritation at flashy descriptions and a hidden jealousy of their fame and success. I could well believe that the philosophers of Galileo's time were envious of his book sales!

The Copernican system that Galileo promoted was not intended as a model that could be used to describe where humanity stands in relation to God. Galileo's position was that Scripture told you "not how the heavens go but how you go to heaven." (Galileo quoted this witty phrase in his *Letter to the Grand Duchess Christina,* attributing it to "a churchman who has been elevated to a very eminent position," most likely his friend Cardinal Caesar Baronius—who died in 1607, well before Galileo's problems began.) This is why some philosophers and theologians had problems with Copernicanism; it didn't work as a metaphysical metaphor, the way the medieval cosmology had.

Furthermore, the Copernican model wasn't really all that much of a mathematical simplification. Even simply describing correctly how the planets, including Earth, go around the sun—the basic idea behind the Copernican system—depends on two further advances that were lacking when Galileo first started to write.

In order for a heliocentric system to work—which is to say, do more than merely compute accurately the positions of the planets as actually observed, but to further understand the real motions of the real planets—you need the planets to be moving in Kepler's elliptical orbits, not the circular orbits that Copernicus assumed. Kepler's first two laws were published in 1609, the same year as the *Sidereus Nuncius*, while the third came out in 1619. Nobody really appreciated them at the time, however. Galileo himself ignored Kepler's ideas and never even acknowledged getting the pamphlets Kepler sent him. It would be a generation later before Edmund Halley, the comet guy, appreciated them and pointed them out to his buddy Isaac Newton.

The other thing you need, which is even more essential, is Newtonian physics to explain *why* planets should move in that fashion. Only then do you have a true union of astronomy and physics. The definitive third edition of Newton's *Principia*, with the clearest explication of how orbits work, only came out in 1726, more than a century after Galileo burst onto the scene.

In fact, before Newton came along, there was another cosmology in play that managed to describe the positions of the planets just as well as Copernicus did without violating the assumptions of Aristotle's physics. In the late 1500s, Tycho Brahe had proposed that the planets orbit the sun, like Copernicus suggested, but the sun (with all the planets in tow) was still in orbit around a stationary Earth. This kept the geometrical benefits of the Copernican system but explained why no one could actually detect any motion of the Earth relative to the distant stars.

While the Jesuits were teaching the Copernican method as a mathematical tool to predict planet positions, the Jesuit priest Giovanni Battista Riccioli published a book, the *New Almagest*, in 1651, that went through a thorough discussion of the scientific arguments for and against the Copernican and Tycho systems. Given the physics of the day (pre-Newton) and the lack of observational evidence you would expect to see if the Earth were moving or spinning (no parallax or aberration seen in the stars' positions, no evidence of what we now call the Coriolis Force), Riccioli concluded that Tycho Brahe's cosmology was perfectly adequate to "save the appearances." "Saving the appearances" was the goal for any mathematical system: to succeed in matching the known motions of the stars and planets. Of course, the Tycho system was also what the Church wanted to hear at the time; a stationary Earth meant you didn't have to reinterpret Scripture. On the other hand, in the map of the moon published in this book (which introduced our modern nomenclature), Riccioli named a crater for himself in the same region of the moon where there are craters named for Copernicus, Kepler, Aristarchus, Galileo, and other famous heliocentrists. Was this a secret message that he, too, preferred the heliocentric system?

By the early 1700s, the Jesuit priest Roger Boscovich started lobbying the Vatican to remove restrictions on the Copernican system. The astronomer James Bradley finally recognized the aberration of starlight, a slight shift in the apparent positions of the stars over the course of a year, in 1729. That provided the first direct evidence that the Earth moves.

Meanwhile, a monument to Galileo was erected in 1734 in the church in Florence where he was buried, and in 1744 Galileo's *Dialogue* was printed (with cautionary notes) in Padua. The Congregation of the Index finally withdrew the decree that prohibited all books teaching the Earth's motion in 1757. The *Dialogue* and a few other books continued to be explicitly included on the Index, however. It wasn't until 1820 that, rather embarrassed they hadn't done so earlier, the Holy Office (see the appendix) officially allowed the heliocentric system to be taught by Catholics "as an established fact." In fact, it had been so taught for at least half a century before then. For example, the ceiling of a Jesuit school in Prague, painted in the 1750s, shows all the stars surrounded by orbiting planets.

Additional support from the Church for Galileo came in 1893, when in the encyclical letter *Providentissimus Deus* Pope Leo XIII put forth a view of the relationship between biblical interpretation and scientific investigation that corresponds to the one advanced by Galileo in his *Letter to the Grand Duchess Christina*. And in 1912, Father Johann Hagen, SJ, a priest at the Vatican Observatory, devised a series of very delicate experiments that used the Coriolis Force to demonstrate the spinning motion of the Earth.

PAUL: I suspect these explanations will raise another question for some people, which we need to address explicitly: Why didn't the Church rise above these concerns? If you're infallible, have a line to the Holy Spirit, and so on, shouldn't you avoid making mistakes like this? If you do make mistakes like this, shouldn't you get out of the Church business?

GUY: The obvious reply is that the Church is a human institution, so, of course, it is open to human mistakes. But this question speaks to a deeper misunderstanding of what the Church is and what it's supposed to be. We hardly have the space (or expertise) here to do justice to that question. But to give a shorthand example, think of the description that the Church is the "repository of faith"; it reminds me of those NASA centers where all the data from old spacecraft missions are kept. The data are real. The way we understand those data changes as we change and learn more about the planets . . . or about God.

PAUL: Another, similar question we often get is, why did it take so long for the Church to pardon Galileo? Well, of course, in fact, the Church didn't "pardon" Galileo. The original trial found him guilty of "suspicion of heresy" (which has a very tricky meaning, hard to appreciate with modern sensibilities), and they ultimately got him on a technicality—that he'd violated his agreement with Cardinal Bellarmine not to publicly advocate the Copernican system. He did seem to ignore that agreement once his friend Urban VIII was elected Pope. The 1992 statement from Pope John Paul II was not exactly a pardon; instead, it was a personal apology to Galileo for what had been done to him. The point was, whether he was guilty or not, the Church never should have gone after Galileo in the first place.

One of the odd things about the whole Galileo story is that the entire tradition of Christian and Catholic theology since Roman times had been for the Church Fathers to treat Scripture on a number of levels, including that of allegory. Only with the rise of Protestantism in the late 1500s and early 1600s—just at the time of Galileo—did Catholic theologians of the time, trying to be "up-to-date" and feeling stung by Protestant attacks for not taking Scripture seriously, start focusing more on only a "literal" understanding . . . with the result that they got it wrong in Galileo's case.

GUY: That's why the Galileo Affair is so often cited by people thinking the Church is antiscience. They don't have much else to cite. It's one of the few examples they can come up with where the Catholic Church made that particular mistake.

Actually, a small group of Roman Catholics today wants to believe that Galileo was wrong and the Church was right to think the planets rotate around the Earth instead of the sun! How can you reason with people who think they are being faithful Catholics by insisting that geocentricism is the absolute truth? I wonder if these people are sincere, or is this a clever bit of theater?

The very idea of insisting "Galileo was right!" or "Galileo was wrong!" highlights a common misconception a lot of people have about what science is and what it does. Science is not a collection of facts that can be labeled "right" or "wrong." It is the way we understand those facts; it is the conversation we have about them, and that understanding, that conversation, is constantly evolving. Of course, we understand the universe in a far more nuanced way than Galileo did four hundred years ago. And I would hope that the next four hundred years see just as much development. That doesn't make Galileo "wrong" any more than we are "wrong" for clinging to our soon-to-be-antiquated twenty-first-century physics.

No scientists actually use Galileo's writings anymore to do their work. (This is very different from a philosopher or an expert in literature, who, of course, will continue to cite Plato or Shakespeare. That's one way we know that the Bible is not a science book—unlike the Bible, science books go out of date.) Regardless, we wouldn't be where we are today without Galileo. As a number of Popes have made clear, Galileo has long been admired as a good Italian and a good Catholic (within the mores of his times, and in spite of the injustice he received from the Church). And he was a great scientist.

GALILEAN ENTANGLEMENT

PAUL: There's something fundamental about the Galileo story, however, that I'd like to develop more. The Galileo Affair shows what can go wrong when philosophy, politics, and personalities intersect. Does that mean that science and religion and politics and culture should not be entangled?

GUY: I guess you could strive for a strict separation of those different threads, but, in fact, it's impossible. We're all human beings, and those threads are entangled within each of us. In fact, often the desires of one thread fuel advances in the others. Our personal curiosity drives our science. Our ability to coordinate our efforts with other people—politics—makes big projects like the moon landings or finding the Higgs boson possible.

PAUL: Furthermore, while the Church professes to uphold certain revealed, eternal truths, those truths can only be expressed in terms of the language of the day, which is supplied—at least in part—by science.

Some people see the movement from Earth-centered, to sun-centered, to galaxy-centered, to the centerless Big Bang as if it were some sort of progressive refutation of Christian beliefs. But there's nothing in fundamental Christian doctrine about Earth being at the center of the universe—no declaration that what makes us special is that everything revolves around us.

GUY: The centrality of Earth in the medieval system was not emblematic of its being a special place, merely of its being the *only* place. The common view back then did not recognize that other planets could be "worlds" that might somehow rival Earth.

In that worldview, the Earth wasn't the center so much as the "bottom" of the universe, just one step above the fires of the inferno. If you want to be technical, the actual "center" of the universe is not Earth, but the lowest circle of hell! In fact, C. S. Lewis, in his book *The Discarded Image,* describes how some people in medieval times took this view of the universe— with hell at the "center" and God at the most extreme distant reaches—as merely the inverse of reality. God must, in fact, be the center, and humans among the farthest removed of his Creation.

PAUL: But Christianity took on-board the idea of an Earth-centered, or Earth-only, cosmos, which was widely accepted by the best scholars of the day, as a natural expression or symbol of the Christian belief that we humans are at the center of God's care and concern. Geocentrism was used that way in biblical commentaries, in iconography, in art, in stained-glass windows. And after a thousand years, geocentrism as a way of expressing this doctrinal belief had become identified with the belief itself. And so Galileo learned, to his dismay, that challenging the geocentric cosmology was perceived by some as tantamount to challenging Christian doctrine itself.

GUY: Of course, Galileo intended no such challenge. But his was a new idea and hard for a lot of folks to get their minds around.

PAUL: When science abandoned the idea that Earth was the only world, Christians had to adjust. They had to find other ways to express and represent their beliefs about how God relates to humanity.

Christians made that adjustment only slowly and with difficulty. It's not that they were stupid. It's that over the centuries the doctrinal belief about the centrality of humanity in God's care and concern had become so deeply and pervasively entangled with its geocentric representation in literature, art, and other realms of culture that it took a long time to tease them apart.

This is a built-in problem for Christianity in relation to science. Christianity is a big social structure that has no choice but to depend on and make use of the language and concepts of science and other sources to give expression to what's important to it. It has no alternative; there is no other language to use. But big social structures have a momentum of their own—they sometimes resist change, and they don't change quickly or easily.

That didn't pose much of a problem prior to five hundred years ago, when science wasn't particularly progressive. But now scientific theories and worldviews are proposed, adopted, refined, explored, found wanting, and discarded over the course of a generation or two or even faster. So if Christianity finds it helpful to make use of a particular scientific theory or concept to symbolize or give expression to some fundamental belief, it needs to be ready to make an adjustment when science moves on.

In fact, some modern theologians have tried to use the concept of entropy to express or symbolize evil, and to use the concept of quantum mechanical indeterminacy to express (or to create space for) the notion of human freedom. These attempts may be theologically interesting, but they haven't caught on—no one is making mosaics or stained-glass windows on those themes.

GUY: I suspect part of the problem is that most people outside of the sciences aren't really familiar with "entropy" or "quantum indeterminacy" in the same gut way that folks in the Middle Ages could appreciate the sphere of the stars, which they could see for themselves on any clear night. Modern physics is too much like abstract art, as you described back in the Chicago Art Institute. You can't engage with it the way you can picture the winds painted on the walls, surrounding us up here in the Tower of the Winds.

On the other hand, I can point out a stunning set of stained-glass windows in the Catholic cathedral in Cairns, Australia. They even incorporate the latest Hubble images to depict the moment of Creation! It's glorious, and it works. It's also science that looks up-to-date today but might seem quaint in five hundred years.

PAUL: And there are cases of scientific theories that seem so well suited to symbolizing and giving expression to a Christian belief that they continue to be used in Christianity even when science has left them behind. That's the case with the doctrine of transubstantiation. Within the thought-world of Aristotelian physics and Thomistic metaphysics, the concept of transubstantiation provides a magnificent, beautiful, and scientifically intelligible way of giving expression to the change that Catholics believe takes place in the Eucharist.

But today's physics has long since left behind the language of Aristotle's physics. Relative to modern physics, it's not that describing something in terms of transubstantiation is wrong—it's that it's *not even* wrong. The Aristotelian language and categories that are used to talk about transubstantiation simply play no role in modern physics—they do not appear. As far as modern physics is concerned, it's not that transubstantiation is a *bad* explanation of change; it's that it's no explanation at all.

But the Church has continued to make active use of the term "transubstantiation" as a preferred way to talk about the Eucharist. Theologians may correct me on this point, but I suppose that, because transubstantiation provides such an extraordinarily beautiful and illuminating account of what happens in the Eucharist, and because it is so deeply embedded in Christian art, literature, and culture, the judgment has been made that abandoning that formulation would hurt more than it would help. It has been decided that in this case it's better to live with a scientific anachronism than try to move on to some other mode of expression. But in that case it becomes incumbent on Christians to make some effort to learn how to swim in the waters of Thomistic metaphysics and Aristotelian physics—at least enough to stay afloat. Otherwise the language of transubstantiation becomes a meaningless token—a magic word used to refer to something that is not understood at all.

Being light on your feet in this way is not easy. Once you express an important theological truth in certain contemporary terms, people get comfortable with those terms and don't want to change. Also, material culture gets built around those terms, and it ain't cheap redoing stained-glass windows!

Augustine foresaw the danger of this sort of thing, as you noted on Day 1 when you quoted his reaction to people trying to use Genesis to challenge "modern" Ptolemaic astronomy. Galileo saw the issue accurately. But most did not. So, if you're going to ask the Church to change a deeply embedded mode of expression because science has changed, you'd better be able to show for sure that it's worth the cost. And this Galileo could not do.

GUY: The fact that Christianity occasionally drags its feet with respect to science doesn't mean that it's opposed to science, or that it wants to vie with science for control of truths about the natural world. But it does mean it can't be expected to encompass all of science, either.

PAUL: Look around this room again. Back when this Tower of the Winds was built, one man might be an expert in geography, art, mechanics, the weather, and meridian-making. That unity of knowledge has been lost . . . in no small part because there's simply too much to know nowadays.

GUY: Oh, and one last irony. It was the Swedish king, Gustavus Adolphus, whose victories in the Thirty Years' War, I argue, may have caused the turmoil in Rome that led to Galileo's troubles. The king's daughter, Christina, became queen of Sweden when Adolphus was killed in battle in 1632. But she converted to Catholicism in 1654. She abdicated the Swedish throne, moved to Rome, and for a while wound up living here, in this very Tower of the Winds!

APPENDIX: A TIME LINE OF KEY EVENTS IN THE GALILEO AFFAIR

1543: Copernicus publishes his book *On the Revolutions of the Heavenly Spheres,* which proposes the idea that the Earth spins and moves around the sun in a series of complicated, nested circular motions. He dies the same year.

Although Copernicus is a Catholic (and, indeed, before he died, was in line to be made a bishop), the person who oversees the publication of his book is Andreas Osiander, a Lutheran in Germany. Osiander inserts a preface in which he asserts (contrary to what is actually stated in the body of the book itself) that the Copernican system should be interpreted *hypothetically.* That shows that even in Copernicus's day there was concern by people like Osiander that a theory such as this should be interpreted solely as a work of mathematics (i.e., solely concerned with correctly predicting the observed positions of the planets, what they called "saving the appearances") rather than as a work of physics (i.e., as being concerned with what is "really" in the heavens). The fact that this concern was raised right from the start poses all the more starkly the question: Why did troubles about this system occur only in Galileo's day, nearly one hundred years later? Why not sooner? For that matter, why not later? Why ever?

1564: Galileo is born in Pisa. Pisa is famous not only for its Tower, but also as the university town affiliated with nearby

Florence, the wealthy city of the Medici family that serves as the capital of Tuscany.

At this time, Italy is divided into a number of independent regions. Among them, the most wealthy and powerful are Venice, ruled as a republic by an oligarchy of wealthy families; Tuscany, ruled as a hereditary dukedom by the Medici family; and the Holy See, a swath of central Italy including Rome that is ruled in civil matters as well as church matters by the Pope, who is elected for life by an international College of Cardinals.

1581–1585: Galileo is enrolled at the University of Pisa to study medicine but spends much of his time studying mathematics with a private tutor. Ultimately, he leaves without a degree.

1588: Galileo is appointed a professor of mathematics in Pisa by Ferdinand I de' Medici (1549–1609), the Grand Duke of Tuscany, despite that fact that he does not have a degree. Ferdinand I is married to a cousin, Christina of Lorraine (1565–1637), the granddaughter of Catherine de' Medici, queen of France. (Beware the profusion of people named Ferdinand. This Ferdinand will eventually have a grandson, also named Ferdinand, who will become the future Grand Duke Ferdinand II of Tuscany; he is not to be confused with the future Holy Roman emperor, also named Ferdinand II.)

During this period Galileo is in correspondence with the Jesuits of the Roman College, the Jesuit school in Rome, and Father Christopher Clavius, SJ, writes letters of recommendation for him.

1590: Ferdinand I and Christina have a son, Cosimo.

1592: Galileo moves north, becoming a professor of mathematics at the University of Padua. Padua is the university town affiliated with Venice, located at the northeastern end of the

Italian peninsula, quite some distance from Florence and very independent of Rome.

1600: Giordano Bruno is burned at the stake by the Holy Office, also known as the Roman Inquisition. How he arrived at that sorry end is a long, complicated story that doesn't make anyone, including Bruno, look good.

Incidentally, the Roman Inquisition is not to be confused with the Spanish Inquisition; its jurisdiction covered only the Italian peninsula and Malta and was mostly concerned with limiting the spread of Protestantism. Cases like Bruno's, ending in the death penalty, were bad but thankfully rare—fewer than 2 percent of all cases they heard. Like the Spanish Inquisition itself, the Roman Inquisition hardly matched the evil reputation we hear about from stories told by the English. Recall that England was fighting wars with Spain at that time—the Spanish Armada had sailed to England just twelve years earlier, in 1588—and the English did everything they could to demonize their enemy. Meanwhile, the English had their Star Chamber proceedings, which—unlike the Inquisition—were held in secret, with no witnesses allowed.

1605–1608: During the summers, Galileo comes back home to Pisa from Padua, where he is now working, and tutors the teenaged Cosimo in mathematics.

1608: Cosimo, eighteen years old, marries Maria Maddalena of Austria, who has just turned nineteen herself. She is the sister of Queen Margaret of Spain and also sister of the future Holy Roman emperor, Ferdinand II-to-be. The following year, Grand Duke Ferdinand dies, and Cosimo (not yet nineteen) becomes the Grand Duke of Tuscany, Cosimo II.

1609: Johannes Kepler publishes his book, *New Astronomy*, in Prague. This is where he proposes the first two of his famous three laws of planetary motion: that planets move in elliptical orbits, not in circles; and that the speed with which they move is related in a rigorous mathematical way to the size of their orbits and where they lie in those orbits. (Fifty years later, Kepler's three laws will help point the way to Newton's Theory of Universal Gravitation.) Kepler sends a copy of *New Astronomy* to Galileo, but it seems that Galileo never understood Kepler's book . . . if he read it at all. Unfortunately, Kepler wrote in a turgid Latin that was nearly impenetrable and that was colored by his idiosyncratic theology.

1609: Galileo makes his first telescope and starts observing the moon, stars, and planets.

1610: In the spring, Galileo publishes his book, *Sidereus Nuncius,* in Venice, in which he summarizes and explains his telescopic observations of the heavens. The title is usually translated as *Starry Messenger,* but the Latin *nuncius* also carries the idea of an "ambassador." (The ambassadors from the Vatican today are still called *nuncios.*) Galileo styles his book, and himself, as an ambassador speaking for the heavens. Galileo dedicates his *Sidereus Nuncius* to Cosimo and names the moons of Jupiter the "Medicean moons."

Grand Duke Cosimo II (now all of twenty) appoints the newly famous Galileo, his one-time math tutor, as chief mathematician and philosopher to the grand duke of Tuscany. Galileo moves back to Florence (leaving behind his girlfriend, and their three children, in Padua. His two daughters eventually enter convents.)

That same year, Cosimo and Maria Maddalena have a son, whom they name Ferdinand (of course).

1610–1616: A series of books is published arguing for and against Galileo's telescopic observations on theological grounds. He has as many defenders as detractors among the theologians.

Often it happens that when someone attacks Galileo personally, soon thereafter they backtrack and apologize. It is clear that Galileo is something of a lightning rod for public attacks, but as the official court philosopher of the powerful Medici family, he is also very well connected and protected.

1611: Galileo travels to Rome, demonstrates his telescope to officials there, and is feted by the Jesuits at the Roman College, including the elderly Father Clavius.

1613: Galileo's *Sunspot Letters* is published in Rome with Church approval, once some biblical references and anti-Aristotelian statements are deleted. Galileo's descriptive account of how sunspots appear, disappear, and move across the face of the sun poses an implicit challenge to the reigning Aristotelian physics, in which heavenly bodies were taken to be perfect and unchanging. Along the way, Galileo attacks the Jesuit priest, Christoph Scheiner, who claimed to have discovered sunspots before Galileo did. (In fact, the Englishman Thomas Harriot saw sunspots with a telescope before either Scheiner or Galileo.) Galileo jealously protects his claims of precedence in discovery and invention.

1615: A consultant to the Holy Office reports that Galileo's views do not contain any significant errors and, indeed, conform essentially to Catholic doctrine. Nonetheless, the Holy Office decides to examine Galileo's *Sunspot Letters*.

Galileo circulates his *Letter to the Grand Duchess Christina*, in which he attempts to explain—ostensibly to the mother of his Grand Duke, but in fact to his religious critics—why the Copernican system is perfectly compatible with the Bible and with the Catholic faith.

1616: At the encouragement of Cardinal Alessandro Orsini, Galileo publishes his *Discourse on the Tides*. In this book, Galileo attempts to use the existence of ocean tides as a physical proof that the Earth spins on its axis. Galileo sees the tides as a kind of sloshing back and forth of the oceans in response to the movement of Earth, akin to the sloshing back and forth that happens when a tank of water is transported in a moving cart. As it happens, Galileo's explanation of the tides is completely wrong. (In his defense, the local geography makes the tides in the Mediterranean around Italy very complicated, unlike the simple twice-a-day tides we are familiar with in the Atlantic or Pacific Oceans.)

A committee of eleven consultants reports to the Holy Office that they are of the unanimous opinion that the proposition that the sun is at the center of the universe and does not move is "philosophically absurd and formally heretical," and the thesis that the Earth moves is "philosophically absurd and theologically erroneous." Note, this is the first time—more than seventy years after its publication—that any official group of theologians makes this judgment about the Copernican system.

Pope Paul V orders Cardinal Bellarmine to warn Galileo to abandon his Copernican views. Bellarmine calls Galileo to his home and gives him the warning, then reports to the Holy Office that Galileo has acquiesced.

The Congregation of the Index (a different group from the Holy Office, charged with identifying heretical books) publishes a decree suspending, until corrected, Copernicus's book *On the Revolutions of the Heavenly Spheres* and similar works. Galileo is not mentioned at all.

In response to rumors that Galileo had been condemned as a heretic, Cardinal Bellarmine writes a declaration on Galileo's behalf, denying these rumors and affirming that Galileo is *not* a heretic.

After his visit with Bellarmine, Galileo writes from Rome to the Tuscan secretary that he had a friendly audience of three-quarters of an hour with Pope Paul V, who received him warmly and reassured him.

1620: Four years after starting on the task, the Congregation of the Index finally issues the promised correction of Copernicus's book *On the Revolutions*. Everyone who owns a copy of that book is instructed to cross out certain offensive lines and write new lines in the margins. Hardly anyone bothers to do so.

1621: Cosimo II, a few months shy of his thirty-first birthday, dies; Ferdinand II (not to be confused with his uncle Ferdinand II of the Holy Roman Empire) inherits the rank of Grand Duke. Since he is only ten years old, two regents are appointed to rule Tuscany and control the family fortunes until he comes of age. These are his grandmother, Christina of Lorraine; and his mother, Maria Maddalena. Grandma is the daughter of a past French king and, thus, a cousin of the current French king; mom's brothers are the king of Spain and the Holy Roman emperor. Thus, a careful political balance is achieved. (At this point the Thirty Years' War has already begun, and it is about to heat up.)

1623: Cardinal Maffeo Barberini is elected Pope Urban VIII. He is an admirer and patron of Galileo. Like Galileo and the Medicis, he is from Tuscany.

Galileo's book *The Assayer* is published in Rome under the auspices of the Academy of the Lynx, a scientific society to which both Galileo and Maffeo Barberini belong. *The Assayer* is famous for containing an early discussion of the philosophy of science. Among other things, Galileo asserts that scientists should base their conclusions on observational evidence rather than on what is read in authoritative ancient books and texts.

This book represents Galileo's final volley in an extended

polemical exchange with the Jesuit scientist Horatio Grassi concerning the location and composition of comets; the topic was current, given that three comets had appeared in 1618. Galileo attacks Grassi and his theory of comets with brilliant and devastating sarcasm. Grassi had observed the comets with a telescope, but Galileo (who never saw the comets) claims that comets are not real objects, just refractions of light in the Earth's atmosphere!

Meanwhile, as a result of the opening set of battles of the Thirty Years' War, Bohemia (today a part of the Czech Republic) is ceded to Ferdinand II, the very Catholic Holy Roman emperor. (The center of the Holy Roman Empire is what we now call Austria.) Family ties are important, so note the following: Ferdinand's sister, Margaret, is married to Philip III, the king of Spain. Ferdinand and Philip are also first cousins, once removed.

The Protestants in Germany are alarmed; Denmark intervenes. The English have been opposed to the Spanish for more than a generation (recall the Armada in 1588) and are keeping an eye on the situation. The French are concerned, as well, with the growth of Spanish power in the center of Europe; they begin to send financial support to the Protestants.

At this point, the Holy Roman emperor commissions Albrecht von Wallenstein to be his general. (Wallenstein, who is a wealthy nobleman, puts up the money to pay for the army in return for being named general.)

1624: Galileo visits Rome for about six weeks and is warmly received by Pope Urban VIII and other Church officials; he is granted papal audiences on six occasions. As he is about to leave Rome, Galileo writes to Prince Cesi, head of the Academy of the Lynx, reporting on his visit. In particular, he mentions having been told the following story by the German cardinal Hohenzollern: it seems that while discussing the subject of Copernicus with the Pope, Hohenzollern told the latter that "all heretics accept his opinion and hold it as most certain, and so one must be very circumspect in arriving at any decision; to this His Holiness replied that the Holy Church had not condemned it, nor was about to condemn it, as heretical, but only as temerarious, and that one should not fear that it could ever be proved to be necessarily true."

In the fall, Galileo begins working on what was to become known as his *Dialogue Concerning the Two Chief World Systems*.

1626: In the Thirty Years' War, Wallenstein defeats the Danes at Dessau Bridge; northern Germany falls under the control of the Holy Roman emperor. England and France withdraw from the war.

1629: The "Edict of Restitution" is issued, whereby Catholics are to be given back territory and estates taken from them as long ago as the 1550s.

1630: May/June: Sweden, led by the military genius King Gustavus Adolphus, intervenes on the side of the Protestants (with French support). Sweden is a potent military power in these days and fervently Protestant.

Meanwhile, Galileo receives a letter from a contact in Rome reporting that Prince Cesi had said, "In the last few days Father Campanella was speaking with His Holiness and told him that he had had the opportunity to convert some German gentlemen to the Catholic faith, and they were very favorably inclined; however, having heard about the prohibition of Copernicus, etc., they had been scandalized, and he had been unable to go further. His Holiness answered, [in] the following exact words: 'It was never our intention [to condemn Copernicus], and if it had been up to us, that decree would not have been issued.' "

Galileo completes work on the *Dialogue Concerning the Two*

Chief World Systems. This book presents the two competing "systems of the world" (what we today call cosmologies), the Earth-centered versus the sun-centered, in the form of a dialogue among three characters. Salviati supports the Copernican view; Simplicio is a follower of Aristotle and the Earth-centered universe; and Sagredo is the neutral observer they are trying to convince. Though the dialogue is set up as if it were a balanced account of the two systems, it is clear even in a casual reading that the Copernican view is strongly favored. But at the end of the book, Galileo includes an avowal, placed in the mouth of Simplicio, that all this should be taken hypothetically.

For about two months Galileo is in Rome to obtain the official Church approval for the *Dialogue* and arrange for its publication by the Academy of the Lynx. On June 26, he leaves Rome with a written endorsement of his book from the Vatican secretary, who was the chief Roman censor. Galileo believes this to be a final approval, pending certain minor additions and finishing touches. But in August Prince Cesi dies, and the Academy of the Lynx is left with no one to replace him for leadership and support. An outbreak of the plague interrupts travel and commerce in Italy. Difficulties and delays develop in executing the practical details of printing the *Dialogue* in accordance with the understanding reached in June between Galileo and the Vatican secretary.

1631: Galileo writes to the Tuscan secretary of state, indicating that he is tired of waiting for the Vatican secretary in regard to the publication of his book. In light of the death of Prince Cesi and complications caused by the plague, he wants to publish his book in Florence. Intense negotiations take place with the Vatican. Ultimately, the Vatican secretary transfers jurisdiction over publication of the book in Florence to the Florentine inquisitor; the Vatican also gives instructions, stemming from the Pope, concerning its title, contents, preface, and ending. (It is clear at this time that the Pope is well familiar with the book, its contents, and its intent.)

In September, the Spanish and Holy Roman Empire forces are defeated by Gustavus Adolphus and the Swedes at Breitenfeld.

On November 1, Maria Maddalena dies. Though at about this time Ferdinand attains his majority, his aging grandma Christina remains the power behind the throne. Christina's policy is generally to align herself with the interests of Pope Urban VIII. However, after her death in 1637, Ferdinand will eventually bankrupt Tuscany fighting wars against the Pope.

late 1631–early 1632: Supporters of Spain in Rome publicly accuse Pope Urban VIII of not supporting their cause; he is even accused of being a secret Protestant! They demand that the Pope send troops to help Spain and that he threaten the French with excommunication if they do not stop fighting Spain. The Spanish know that Urban VIII has ties to the French, who helped get him elected Pope.

1632: After more than a year of negotiations between Rome and Florence, the *Dialogue* is finally published in Florence. It is received there with great enthusiasm.

In March, Spanish forces are defeated at Lech. The Spanish ambassador to Rome, Cardinal Gasparo Borgia, attacks the Pope in a consistory meeting. When his speech is stopped by the Pope, he has the rest of his complaint printed and published.

However, the French cardinal Richelieu threatens to break with the Roman Church if the Pope acts against France. (Yes, that Cardinal Richelieu. These are the times later fictionalized in *The Three Musketeers* . . . which, like many an adventure set in historical times, is not exactly an accurate account of history!)

In May, Catholic Munich falls to the Swedes.

In November the Swedes win again, at Lützen, but King Gustavus Adolphus is killed.

Meanwhile, a number of issues have been arising for Galileo in Rome. Pope Urban VIII prohibits the printer from further distributing the *Dialogue* and appoints a special commission to investigate the matter. The commission files its report, and the Pope decides to forward the case to the Holy Office. At a meeting of the Holy Office, with the Pope presiding, it is decided to summon Galileo to Rome. On hearing the news, Galileo signs a written statement promising to obey.

Galileo then writes to Cardinal Francesco Barberini, the Pope's nephew, who had been tutored by Galileo ten years earlier while studying at the University of Pisa. Galileo asks that the trial be transferred to Florence, or that he be allowed to answer the charges in writing. At a meeting of the Holy Office, the Pope presiding, Galileo's plea is refused. Three Florentine physicians sign a statement that they have examined Galileo, arguing that a trip to Rome would be dangerous to his health (Galileo is sixty-eight years old and partially blind). At a meeting of the Holy Office, the Pope presiding, Galileo's medical excuses are dismissed. Either Galileo must come to Rome on his own, or he will be arrested and brought to Rome. Given that Tuscany is ruled by the Medici, not by Rome, this threat carries certain political overtones.

1633: Early in the year, Wallenstein, the general hired by the emperor Ferdinand II, plots for power against Emperor Ferdinand.

Galileo makes his will and arrives in Rome. At first he stays at the Tuscan embassy, where he is ordered not to socialize and to remain in seclusion. Cardinal Francesco Barberini is now personally handling the case; he notifies the Tuscan ambassador that for the upcoming Holy Office examination, Galileo will have to move to the palace of the Holy Office.

Finally, Galileo is formally interrogated by the Holy Office and is kept at the Holy Office headquarters, staying in the chief prosecutor's apartment. Three Holy Office consultants file their reports about Galileo's *Dialogue*.

The commissary general of the Holy Office reports to Cardinal Francesco Barberini that he has arranged a deal whereby Galileo will admit some wrongdoing and then will be treated with leniency. Galileo signs a statement in which he admits to some inappropriateness but to no malicious intent in connection with the writing of the *Dialogue*. He is allowed to return to the Tuscan embassy, with orders not to discuss the case with anyone, and to return to the Holy Office headquarters when summoned again.

When Galileo appears before the Holy Office again, he makes a formal presentation of his defense, including Bellarmine's certificate from 1616 that he is not a heretic.

Once more the Holy Office meets, Pope Urban VIII presiding. The Pope refuses the deal. He decides that Galileo must be examined on his intentions, "even to the extent of torture" (no such torture ever takes place); that Galileo is to abjure the heresy of which he is suspected before the full Holy Congregation, promising never again to speak or write that the Earth moves or that the sun does not move; that the *Dialogue* be prohibited; and that, to be used as an example, his sentence be sent to apostolic nuncios and inquisitors to be read in public to gatherings of professors of mathematics. The Holy Office then examines Galileo with the formal threat of torture to determine his intention.

On June 22, 1633, Galileo is found guilty of "vehement suspicion of heresy" and sentenced, with only seven of the ten cardinal inquisitors signing the sentence. His punishment involves an abjuration, prohibition of the *Dialogue*, formal arrest at the pleasure of the Holy Office, and some religious penances. Galileo recites in public a formal abjuration: "I abjure, curse, and detest the errors and heresies [of the Co-pernican system] and in general each and every other error, heresy, and sect contrary to Holy Church." The actual here-sies in question are never specified. (The legend that he then whispered, *"eppur si muove"*—"still it moves"—dates from a hundred years after the event.)

After the trial, Galileo remains under house arrest at the Tuscan embassy. The Pope refuses Galileo's petition to be al-lowed to go back to Florence; but, as a step in that direction, he permits him to stay under conditions of house arrest at the residence of the archbishop of Siena (in Tuscany). Gali-leo arrives in Siena in July.

In December, Pope Urban VIII grants Galileo's petition to return to his own villa in Arcetri, near Florence, though the conditions of house arrest will remain in effect.

1634: In February, Wallenstein is killed by Ferdinand's agents.

In September, the Swedes and their allies are defeated at Nördlingen; Emperor Ferdinand II makes peace in Ger-many, abandoning the Edict of Restitution. However, as the Swedes withdraw, the French enter, and the war continues for another fourteen years . . .

1638: Five years after his trial, Galileo publishes (in Holland) his book *Discourses and Mathematical Demonstrations on Two New Sciences Pertaining to Mechanics and to Local Motion.* This is where Galileo lays out his understanding of the physics of motion, an important foundation for Newton's three Laws of Motion forty years later.

1642: Galileo dies at Arcetri. He is buried at the Basilica of Santa Croce in Florence.

1648: The Treaty of Westphalia ends the Thirty Years' War.

DAY 4: What Was the Star of Bethlehem?

SETTING: VATICAN OBSERVATORY TELESCOPES ATOP
THE PAPAL SUMMER PALACE, CASTEL GANDOLFO

A STAR WITH A TALE

PAUL: Today we are visiting what was, until recently, the headquarters of the Vatican Observatory. We're on the rooftop terrace over the Pope's summer residence in the small town of Castel Gandolfo, in the Alban Hills just south of Rome. Our new observatory headquarters are quite close by, in a remodeled monastery less than a mile from here. That's where you and I and the other Vatican astronomers live and work now. But for seventy years they lived and worked right here. Though we don't actually live here anymore, we can still come and use these two venerable telescopes, which are located under beautiful wooden domes, atop the Pope's vacation house.

GUY: I love coming up here to use the telescopes. But also I just love the view from up here. I love walking past the vintage 1930s telescope domes to the edge of the roof to look over the railing down at Lake Albano, in the gorgeous volcanic crater below us. I like looking out over Rome and over the Mediterranean, far away. And I like looking across Lake Albano at the little hillside town of Rocca di Papa, on the shoulders of Monte Cavo, the extinct volcanic cone that rises up about three thousand feet. It's breathtaking.

PAUL: I'm glad you like the view. But I'm hungry for a plate of *pasta caccio e pepe* at our favorite little restaurant, just across the piazza in Castel Gandolfo, Sor Capanna. It's December, the sun is setting, and that damp wind is cutting right through me. Why don't we go get some food?

GUY: Hold your horses! Dusk in December is the best time to come up here! Look across Lake Albano at Rocca di Papa. Do you see the huge "comet" outlined in electric lights, erected on the hillside above the village? They put it up every Christmas to represent the Star of Bethlehem. Neat, huh?

PAUL: It beats looking at a giant glowing Santa!

GUY: An illuminated comet is *the* emblem of Christmas in Italy, much more so than in America. You see it everywhere . . . comets over the doors of churches, on walls, above crèches. It's almost always depicted as a big star with a long, curved tail, lit up with lots of bling.

PAUL: The comet above Rocca di Papa is pretty impressive. It must be, what, three miles away? And yet we can see it clear as day. It must be visible for miles around.

But wait a minute. What *was* the Star of Bethlehem? It wasn't actually a comet, was it?

GUY: That's a question we always get asked at the observatory around Christmas. What was the Star of Bethlehem?

In fact, a few years ago, a reporter for the British newspaper *The Independent* wrote an article stating with great authority—and without actually bothering to get in touch with us—that the chief business of the Vatican Observatory was to determine the precise nature of the Star of Bethlehem! He obviously hadn't done his research. Years earlier, the *Chicago Tribune* had asserted, with equal authority, that our real task was to cast the Pope's horoscopes!

PAUL: I can see why people would want to ask Vatican astronomers about the Star of Bethlehem, especially during the Christmas season. After all, the Star has connections with both faith and science. Since it appears in the Bible, people of faith want to know about it: Was there really a Star? Was it a miracle, or can it be explained by science? What does the Star mean—what does it signify? And since the Star would have been an observable phenomenon, scientists are interested in figuring out whether there really was a Star and what sort of thing it might have been.

GUY: As astronomers, we can ask, was there some sort of highly unusual event in the sky around the time Jesus was born? And if so, what might it have been? If there really was something unusual in the sky, it could have been something supernatural or miraculous, something beyond the scope of science to explain. Or it could have been some sort of natural phenomenon, something that can be explained scientifically. In that case, the only thing miraculous about it would be its very special timing and location—a "divine coincidence" allowing it to function as a special sign of the birth of Christ.

When it comes to data concerning the Star, we don't have much to go on, just these verses from the opening of Matthew's Gospel:

> In the time of King Herod, after Jesus was born in Bethlehem of Judea, wise men from the East came to Jerusalem, asking,

"Where is the child who has been born king of the Jews? For we observed his star at its rising and have come to pay him homage."

When King Herod heard this, he was frightened, and all Jerusalem with him; and calling together all the chief priests and scribes of the people, he inquired of them where the Messiah was to be born.

They told him, "In Bethlehem of Judea; for so it has been written by the prophet: 'And you, Bethlehem, in the land of Judah, are by no means least among the rulers of Judah; for from you shall come a ruler who is to shepherd my people Israel.'"

Then Herod secretly called for the wise men and learned from them the exact time the star had appeared. Then he sent them to Bethlehem, saying, "Go and search diligently for the child; and when you have found him, bring me word so that I may also go and pay him homage."

When they had heard the king, they set out; and there, ahead of them, went the star that they had seen at its rising, until it stopped over the place where the child was.

When they saw that the star had stopped, they were over-whelmed with joy. On entering the house, they saw the child with Mary his mother; and they knelt down and paid him homage. Then, opening their treasure chests, they offered him gifts of gold, frankincense, and myrrh.

And having been warned in a dream not to return to Herod, they left for their own country by another road.

And that's it. Twelve verses. That's all the data we have to work with.

PAUL: Maybe in the end we won't be able to know with certainty whether or not there was something unusual in the sky at the time of Christ's birth. We can return to that question later. But science should at least be able to say what sort of thing the Star might have been, if there was a Star.

So, how about it, Guy? What sorts of astronomical phenomena are possible candidates for explaining the Star of Bethlehem? How would you, as a scientist, go about trying to explain those twelve verses from Matthew, if (for the sake of argument) we accept them as a historically accurate report of what was seen in the sky?

AN ASTRONOMICAL EVENT?

GUY: Well, the Star, if there was one, must have been something unusual and extraordinary in appearance. But the thing about the stars is that, with certain exceptions, they are orderly and pre-dictable. The same constellations appear during the same seasons every year, and any change in their positions is so slow as to be un-detectable by the human eye during one person's lifetime. So we're looking for something that would be a big exception.

One big exception is stars that suddenly go "nova"—or, more impressively, "supernova." A nova, a faint star briefly becoming a bright star before fading back to its former obscurity, is notable if you know the stars well, but otherwise it might go unnoticed. These appear fairly often, however; probably one is visible to the naked eye every ten years or so. That hardly seems spectacular enough to announce the coming of the Messiah. On those grounds, we can rule out the idea that the Star of Bethlehem was a regular nova.

Supernovas are spectacular and unusual lights in the sky; if close enough to us, they can be seen even in the daytime. This was how Arthur C. Clarke explained the Star of Bethlehem in his famous short story "The Star." A supernova would be a real attention-getter. And supernovas are far more rare than mere novas. The last supernova in our galaxy visible to the naked eye was observed in 1604, more than four hundred years ago. (Another naked-eye supernova occurred in our near-neighbor galaxy, the Large Magellanic Cloud, in 1987.)

However, supernovas leave behind remnants. For example, Chinese and Arabic astronomers in 1054 reported a bright light in the sky in the constellation Taurus that matches what we recognize today as a supernova. If you look at that spot in the sky today, even with a small telescope you can see the Crab Nebula, the leftover ashes of that stellar explosion. More tellingly, supernova remnants like the Crab are quite prominent sources of radio emissions, which we attribute to the rapidly spinning core of the burned-out star. So for radio astronomers, they are hard to miss. We can figure out how much a supernova nebula should have dissipated and spread out over a period of two thousand years; we've been watching the Crab Nebula spread out for hundreds of years, ever since it was first seen in a telescope (by John Bevis, in 1731). But there are no independent reports of a supernova around the time Jesus was born. And there are no unaccounted-for supernova remains from two thousand years ago. So it seems to me that astronomy can pretty much rule out the idea that the Star of Bethlehem was a supernova.

What else might the Star have been? Well, comets are another spectacular and startling sign in the sky. As we've seen here in Italy, they're a favorite candidate simply for being so intriguing and beautiful. There are a handful of comets, like Halley's Comet, whose regular returns can be predicted. But for the most part comets orbit so far from us that we don't even know they're there until they make their once-in-a-million-years passage close to the sun; then they move out of our range of view again for another million years. So we can't rule out the possibility that there was a bright comet present at just the right moment, two thousand years ago—and we wouldn't expect to be able to see that comet today. But in ancient times comets were universally interpreted as signs of doom. So it's a little hard to see how anyone would see a comet as signifying an event as joyful as the birth of a king or the Messiah.

So, if we rule out supernovas and comets, what else might the

Star have been? Nowadays most attempts at giving a possible scientific explanation for the Star of Bethlehem involve looking for interesting or unusual conjunctions of the planets.

In the early seventeenth century, Johannes Kepler—armed for the first time with accurate records of the positions of the planets (which he got from Tycho Brahe's observations) and with his new theory for planetary motions in ellipses about the sun—tried to play this game by making tedious calculations of planetary positions for the time of Christ. Kepler had actually seen the supernova of 1604, but he had no idea what caused it. He proposed that the supernova and other "new stars" were somehow the result of planetary conjunctions. He figured out that there was a particularly interesting series of conjunctions of Jupiter and Saturn that occurred in 7 BC.

PAUL: Nowadays, anyone with a personal computer can select a date—say, December 25, 1 BC—and plug it into a planetarium software package, to see where in the sky the planets would have been on that night. Of course, it was only in the fourth century after Christ that people started celebrating Christmas in December. No one really knows Jesus's exact birth date, and it's not a matter of Catholic doctrine or belief that Jesus was born on December 25. And in any event, historians now realize that when Dionysius Exiguus started the tradition of counting years from the birth of Christ, in AD 525, he miscounted slightly.

GUY: If you're going to try to explain the Star of Bethlehem as a conjunction of planets, the trick is to find a solution that is consistent with the temporal setting of the Gospel nativity stories (a year while Herod was still alive and king, and probably in the spring, when shepherds would be most likely to be out at night tending their flocks); consistent with the description of the Star (presumably something to do with its rising); consistent with an explanation of how it would indicate the birth of a king, and how Judea would be indicated as the location of this birth; and, finally, consistent with the apparent fact that only astrologers were wise to this event. That's a lot of constraints!

One elegant modern theory I particularly like comes from a book by the astronomer Michael Molnar, *Star of Bethlehem: The Legacy of the Magi*.

Molnar starts with three things in the story that puzzled him. First, any celestial phenomenon, be it comet or nova or alignment of planets, would have the whole world spinning underneath it as it rises and sets. How could something in the sky lead the astrologers specifically to any one particular location, like Bethlehem?

Second, comets and supernovae are rather frightening phenomena, putting an unpredictable (and unpredicted) light in a sky that normally is most comforting for being so beautifully and soothingly regular in its motions. And, as we have noted, comets and the like have traditionally been omens of bad fortune. What kind of celestial event would signify the birth of a king to the astrologers of that time?

And, finally, if it was a spectacular celestial event, why didn't everybody see it? Why did it take wise men (and foreigners, at that) to notice it?

Molnar answers all three in terms of the astrology of the day. Each region of the world was associated with a different zodiac sign, he says, with the region around Jerusalem being connected with Aries the Ram. (He brings various bits of evidence, such as local coins featuring leaping rams, to bolster this point.)

He uses the well-known astrological alignment of planets and the sun at the birth of Caesar Augustus, which Augustus claimed as evidence of his own destined royalty, to argue that such an alignment of planets was well established at that time as a sign of kingship. Molnar points out that the key to the alignment was having the planets rise with the sun—what is called a *"heliacal rising."* Such conjunctions of planets and sun were long considered significant. For example, the heliacal rising in August of the bright star Sirius of the constellation Canis Major, the Big Dog, was a sign to the Egyptians to get ready for the flooding of the Nile. (And it's the origin of our phrase "dog days" for the heat of August.)

But note—only an astrologer calculating the planetary positions on paper would know when heliacal risings occurred. Nobody would be able to actually see the planets rising, since the presence of the sun among the planets makes it impossible to actually observe them.

Molnar then searches the years around the time of Jesus's birth for such a conjunction of all the planets: one with Venus, Mercury, Mars, Jupiter, and Saturn all rising with the sun and the new moon, similar to what was used by Augustus to support his own royal birth. (Remember our conversation during our first day at the Art Institute of Chicago? We talked about how in ancient cosmology all those planets were seen as being the home of the "thrones, dominions, powers"—planetary intelligences, or demons, or deities of the old cosmology.)

Molnar finds that just such a grouping of planets occurred in the constellation Aries in late March of 6 BC, and again in April. And that is the year, he tells us, that historians have identified as the most likely year of Jesus's birth. (Two years later, in spring of 4 BC, the planets are a bit more spread out but at this time they all trail behind the sun at its rising—strikingly reminiscent of how Saint Paul describes that Jesus "has stripped the sovereignties and the ruling forces and paraded them in public behind him in his triumphal procession!" in the passage from Colossians 2:15 that we quoted on Day 1.)

It sounds quite convincing. It's all quite neat. And, indeed, it's rather startling to realize that such an event really did occur in the sky about the time Jesus may well have been born. If your home computer has a planetarium program, you can look it up for yourself.

PAUL: But was this really what Matthew was talking about?

GUY: Good question. We don't even know if Matthew really meant to describe an actual astronomical event at all. And, for that matter, Molnar's theory is hardly the only astronomical explanation that's been suggested.

In his book *The Christmas Star* (Griffith Observatory, 1987), John Mosley of the Griffith Observatory in Los Angeles argues that the conjunction of Jupiter and Venus in August of 3 BC was the beginning of a series of close conjunctions that would have been of interest to any astrologer. The August conjunction took place in the constellation of Leo, which Mosley informs us was the astrological sign of Judah (recall the "Lion of Judah" mentioned in Genesis 49:9). Later, Jupiter (king of the gods) made three close passes by the brightest star in Leo, Regulus (the star of kings). Then, on June 17, 2 BC, Jupiter and Venus encountered each other again, so close that they would have appeared to the naked eye to actually be merged into one "star." Recall, Matthew talks about the Magi looking at a star, *one* star, not at a grouping of many stars. Such close conjunctions occur less than once a century; the last one, Mosley calculates, occurred in 1818, and the next will occur in 2065. And, of course, he points out the historical evidence that tells him the most likely years for the birth of Christ were 3 or 2 BC. (Other historians dispute this, citing evidence that Herod was dead by then.)

But wait . . . as Kepler had originally suggested, there was also a series of spectacular conjunctions of Jupiter and other planets in the year 7 BC. In his book *The Bible and Astronomy*, Father Gustav Teres, SJ (who worked at the Vatican Observatory in the 1980s and 1990s), points out that on November 12 of that year, Jupiter and Saturn met in the constellation of Pisces the Fish, a constellation that we are told was associated with Israel (and appropriate for One who would send out Fishers of Men), in a location that may have been lined up with a thin cone of faint light called the Zodiacal light. That's the faint reflection of sunlight off dust in the asteroid belt, almost impossible to see nowadays with all our artificial lighting, except in dark, remote places like the Oklahoma panhandle—which is where

I was the only time I ever saw it. But an observer of the dark skies of ancient Palestine would have seen a beam of light from these conjoined stars that night, sweeping down from them to the horizon! And we are told that 7 BC is also the year (you guessed it) that he tells us historians have identified as the most likely year of Jesus's birth.

Those are only three of countless different theories you can find in various books and videos or on the Internet. Type "Star of Bethlehem" into Amazon.com and you'll find more than four hundred books listed. It's as popular as Galileo. Each book proclaims that it has the one true explanation for the Star. And each gives vigorous arguments why all the other theories are obviously no good.

PAUL: So, it's not as though modern astronomy has shown that the story in Matthew about the Star of Bethlehem cannot be true. Just the opposite: modern astronomy makes it possible to come up with a surfeit of possible explanations for the Star.

GUY: Even a brief perusal of these different explanations demonstrates pretty convincingly that, if nothing else, the (usually unnamed) "historians" they cite are as uncertain as the astronomers themselves as to which constellation really signified a king of the Jews, and what year is the most likely year of Jesus's birth. It is not utterly beyond the realm of possibility that it actually did happen on December 25, 1 BC!

No one knows enough to argue convincingly which of these theories would provide the best scientific explanation for the Star, if there was a Star. Astronomy, which we had looked to as a source of objective truth, turns out to fail us—not for having no answer, but for having too many possible answers.

PAUL: That's not failure. That's just science doing its main job: to distinguish between what sorts of things can happen in nature and what sorts of things cannot happen.

A PIOUS STORY?

PAUL: OK, you've listed possible scientific explanations for the Star—and it turns out there are quite a few. Should that matter, from the perspective of Christian faith? On one hand, astronomy can show that there were some remarkable events visible in the sky around the time of Christ's birth. That could have the effect of bolstering faith and increasing trust in the Gospel of Matthew. On the other hand, astronomy can't say which (if any) of those events should be considered the Star of Bethlehem. And astronomy indicates that similarly remarkable events have been visible in the sky at various times throughout history. That could have the effect of weakening faith and diminishing trust in the Bible.

GUY: I wouldn't recommend adopting either of those extreme conclusions. But certainly there are elements of both attitudes, credulity and skepticism, in those who seek scientific explanations for the Star of Bethlehem.

PAUL: Suppose that astronomy could show that, out of all the various astronomical events you have listed, one of them was far more remarkable than the rest. Even then, it would remain an open question whether Matthew's story in the Bible was inspired by that particular event. And even if it was, we could still wonder why Matthew chose to record it. What did the story of the Star mean to Matthew—why was it important to him? Presumably the story of the Star, as situated in the Gospel of Matthew, is meant to reveal something about God and about the Kingdom of God. Yes?

GUY: So, now we're stepping back from the question of whether there really was a Star or what sort of thing the Star might have been. Instead, we're asking: What's the point of including a story about a remarkable Star in Matthew's Gospel? What does that story mean, in the context of the Gospel and in the context of Christian faith?

PAUL: You called the star "remarkable." Something can be remarkable scientifically: maybe it's highly unusual, or maybe it seems to violate accepted scientific laws. Or something can be remarkable for reasons altogether different, which have nothing to do with science.

GUY: And, by the way, what counts as "remarkable" scientifically changes over time. An astronomical event that is remarkable and exciting in one context or age might be completely unremarkable in another. Years ago, when I lived in Michigan on the shores of Lake Huron, a neighbor told us a story about her three-year-old grandson, who had been visiting for Christmas. One morning the two of them watched through their living room window as the sun rose over the lake; it's a remarkable sight, the red and golden light of the sky reflected off clouds and water. But the next morning, she heard a shout from the living room: "Grandma!" cried her grandson. "Come quick! It's doing it again!"

PAUL: Even if no scientifically remarkable events occurred in the skies around the time of the birth of Jesus, that wouldn't necessarily make Matthew into some sort of liar. When it comes to stories, we evaluate truth and falsity on multiple levels. You could ask whether a fable is true—for example, the fable of the tortoise and the hare. In that case you're not asking whether the events recounted in the story really did occur; you don't need a professor of zoology to tell you that tortoises and hares don't actually stage races. Instead, you're inquiring about the truth of the moral at the heart of the story, which is that slow and steady wins the race. Zoology doesn't enter into it.

Scripture is written in many different genres. There are historical accounts, mythic stories, morality tales, poetry, and more. At least from the perspective of the Catholic faith and mainline Protestantism, it's a mistake to try to read every line of Scripture as if it were intended by its author to be understood as a literal account of historical events that actually happened. That's just not how many parts of Scripture were meant to be read.

Fundamentalists of both types, religious and scientific, would like to insist that every line of Scripture must be interpreted literally. Religious fundamentalists insist that every line of Scripture must be literally true, and that we should refuse to accept anything science says that is at odds with the literal sense of Scripture. Scientific fundamentalists also insist that the Bible be interpreted literally; however, from this they conclude that the Bible should be rejected, since portions of it, when interpreted literally, are at odds with modern science.

Look, I know it's not easy or obvious to figure out which portions of the Bible we should interpret literally and which portions we should interpret in other ways. But the fact that it's not easy doesn't mean it's not worth doing; after all, science itself isn't always easy. And, of course, we have to guard against the temptation to interpret the Bible according to our own whim or taste, much as scientists have to guard against the temptation to cherry-pick the data.

Over the centuries, standards have changed in Scripture interpretation. At the time of the Protestant Reformation, there started to be a greater (but not exclusive) focus on the plain, literal sense of Scripture. But it was only in nineteenth-century America that full-fledged modern biblical fundamentalism picked up a serious head of steam.

Just as standards of interpretation have shifted in religion, so, too, they have shifted in science itself: there has been change over time in what it means in science for something to be real or true.

For example, in medieval commentaries on classical scientific texts, you can find extended discussions and arguments about the true natures of things like unicorns and manticores. Unicorns and manticores were taken to be *real*. But this was not because people were stupid or naïve or credulous, and not because people thought they knew for sure that such things existed in fact. Instead, at that time what made something count as *real* wasn't mainly that it existed in fact.

For educated people, what made something count as *real* was that it was mentioned in an ancient, authoritative text, as in Aristotle or Galen. That was the criterion of reality—that is what counted, even more than existence in fact. It was only with the rise of modern science, in the seventeenth century, that the meaning of "real" shifted in the direction of how we understand it now. When we say something is real, we mean that it exists in fact and not merely in an authoritative text.

GUY: In some ways that's not all that different from our modern usage. If you say the words "Santa Claus," everyone will know what you mean and what it signifies, even if no one can find a guy in a red suit at the North Pole. In that sense, "Santa Claus" and all it stands for is real.

PAUL: Yes, Virginia.

GUY: And there's also the issue of how different people might interpret the same words. Think of the difference between an astronomical comet and the electric-light "comet" on the hillside across the lake. Most astronomers, when they hear the word "comet," think of a fuzzy bit of light you might see through binoculars or a telescope; most Italians think of something like that remarkable Christmas display over there, across the lake.

PAUL: I agree, the view of that "comet" from up here is remarkable. But I'm cold. Can we go somewhere out of the wind, at least?

GUY: Sure. Let's head up those stairs under that big plaque, and we can go inside the dome of the double-astrograph telescope.

PAUL: The plaque displays the motto of the Vatican Observatory, given to us by Pope Pius XI: *"Deum creatorem venite adoremus."* Just from knowing Christmas carols you can translate that one: "Come let us adore God, the Creator."

OK, now we're out of the wind. But it's still pretty chilly in here. That big old wooden dome keeps some of the heat in, and these solid concrete walls help, but the huge battleship-gray metal structure holding up the telescope makes me feel cold just looking at it.

I never quite understood the design of this double-astrograph telescope. It looks like a giant pair of binoculars. Why are there two scopes, side by side, on one mounting?

GUY: It's a kind of design nobody uses anymore. The idea may have been to save money—put two different telescopes, designed for different kinds of observations, under one dome. In those days, when optics weren't as good as they are today, you needed to build separate telescopes, each optimized for a given task. On the other hand, one story I've heard is that it was designed to photograph the same object in two different wavelengths, two sets of colors. Remember, back then, even black-and-white film was sensitive only to certain wavelengths of light, either the blue end of the spectrum or the red end. Modern "panchromatic" film, sensitive to all visible wavelengths, was invented a few years after this telescope was built. Remember the old Kodak films that had the word "pan" in their names? Until that "pan" film came along, they needed to use one telescope to photograph a star field with red-sensitive film, while the other was used at the same time for blue-sensitive film.

PAUL: So this telescope was designed to address a particular problem that no longer exists!

GUY: We have a similar difficulty when we try to read and understand ancient literature, like the Bible. Much of ancient literature was created to address a problem that for the most part we don't have anymore: how to communicate something, via the written word, to people who can't read?

Nowadays almost everyone can read; and since the invention of the printing press, you're never too far from a book if there's some idea or fact you need to look up. (That was true even before the Internet came along.) But in ancient times, people who could read were rare, and books themselves were equally rare. Most people could only learn from what they heard, usually what was read to them during a church service by the one guy in the village who could read, the local rabbi or parish priest.

That means that books were written with "hooks" that would make the message memorable, something you wouldn't forget even if you only heard it once a year (or less). The important information and truths that Scripture conveys—about who we are and how we should live—were often communicated in the form of stories. You can remember a story even after hearing it just once. Even nowadays, when you leave church on Sunday, you'll probably already have forgotten the theology in Paul's epistle, but you'll probably remember which Gospel story you heard.

PAUL: So Scripture was made to be user-friendly, by means of using various genres! Your smartphone is "user-friendly"; it's a little black box that does what you want, and you don't need to ask how or why. It's based on all kinds of modern electronics and the fundamental physics of electromagnetism. But most people don't know all that physics, and they don't need to know it. Stories were "user-friendly communications technology" in an age when people couldn't read. Stories served to make it easier to communicate and remember Christian faith and doctrine.

GUY: But that means there's a danger of confusing the literary details of a story with what it's supposed to communicate—just like I can get so fascinated by playing with my smartphone that I forget I'm supposed to be calling someone! Probably preliterate populations were well-practiced at separating the message from the medium. But we moderns aren't always so good at that.

Anyway, we shouldn't be surprised when Scripture gives us stories instead of straight facts. Storytelling was one of the necessary techniques commonly used when the Bible was written. (It's not just in the Bible; you'll also find it in Greek mythology and almost every culture's origin story.)

PAUL: It's only in the modern period, the period of modern science, that people start trying to distinguish cleanly between the medium and the message. It's not so clear to me that that's really possible to do all the time. But I think that's the presupposition that biblical fundamentalists bring to their reading of Scripture. They presume that it's possible to get the "message" of Scripture without taking into account the nature of the "medium"—the various genres.

GUY: Another reason for using stories to convey important truths is that sometimes essential elements of the truth simply can't be conveyed in a list of mere facts. There is more to reality than that which can be weighed and measured or videotaped or recorded in a book. Ask anyone who has ever owned a classic 1966 Volkswagen Beetle: there was more to that car than what you would find listed in the Chilton auto repair manual.

PAUL: So, let's get back to our main topic. Was the Star of Bethlehem *real* or not?

GUY: Well, one thing we know for sure. Whether or not there was a Star or Magi, there is a story about the Star and the Magi. The very fact of the existence of that story, regardless of whether it could be confirmed by a time-traveler with a video camera, raises interesting questions.

One possible way to read those verses in Matthew's Gospel is as a symbolic tale, one written to signify that Jesus was as much a king as any secular ruler. Caesar Augustus, after all, used astrological signs to shore up his legitimacy as Rome's emperor; why not do the same for Jesus? Or it could have been written to foreshadow how the message of Jesus would find fertile ground among the Gentiles.

But if the story is "merely" symbolic, just a "pious tale," then there are some nagging questions.

Were such "pious tales" common in the culture in which Matthew's Gospel was written, to the point that anyone reading this story in that day would have understood how it was meant to be taken? And would the author of that Gospel likely have been subtle enough to write such a symbolic passage?

Plenty was written in the Christian milieu in the century following the Gospels, including a lot that is clearly meant symbolically—indeed, the Church Fathers by the second and third century after Christ tended to read all of the Old Testament as if it were an allegorical prophecy of the coming of Christ. But that's a hundred years or more after the Gospels were written. There isn't a whole lot of other literature that came from specifically the same time and place as the Gospels, except for the Gospels themselves. Interpreting something written in AD 75 by assuming it's like something written in AD 275 would be like interpreting the nineteenth-century Lord Byron's poetry as if it were rap: there is just enough similarity that you could be seriously misled.

PAUL: Incidentally, a quick Google search says that there's a modern rap artist who calls himself Lord Byron. No doubt, that will confuse scholars two thousand years from now.

GUY: Maybe outreach to the Gentiles was what Matthew was actually trying to accomplish with his Gospel. Or maybe he was trying to speak more to a Jewish audience. You can find Scripture scholars who argue both sides. In that dispute, the Magi story is cited as evidence that Matthew's Gospel was written for Gentiles like the Magi.

But if this is a pious story invented by Matthew to show wise Gentiles accepting Jesus, then why would he have his "wise men" coming from the *East*? And why would he deliberately tie them to the "Magi"?

The Magi (the term for the wise men used in the original Greek version of Matthew's Gospel) were a class of priest-astrologers who predated not only classical Roman and Greek culture, but also the ancient civilizations of the Medes and the Persians; they were around even before the eastern religion of Zoroastrianism. While there were still traders coming to Palestine from that region at the time of the Gospels, by then Greece and Rome were the center of culture and wisdom—and they were also the home of the Gentiles who were coming into Matthew's church. So if the idea was to create a story showing Jesus as a figure appealing even to the non-Jews, you would think that Matthew would have his wise men coming from Greece or Rome. (But, of course, it's possible that "the Magi" and "the East" had a significance to readers at that time that is lost to us today.)

Another curious part of the story is that *astrology* would be used to announce the coming of the Christ. Astrology held a most unfavorable position in ancient Jewish culture. The Jews condemned the idea that the stars had divine power, or that one could prophesy by the stars. For example, see what Deuteronomy (4:19), or Isaiah (47:10–14), or the book of Wisdom (chapters 7 and 13) have to say about this. Though knowledge of the stars and the seasons is wisdom that comes from God, saying that the stars control our personality or fate was seen as a denial of personal freedom and responsibility. It's also far too close to the idea of worshiping the stars instead of the one true God.

But that doesn't mean the Jews didn't accept the cosmology of their times. You can find mosaics illustrating the zodiac in ancient synagogues. The best science of their day taught them that the complex and bizarre changes that occur in human affairs could find parallels in the complex and at times bizarre motions of the planets. Indeed, the classic Yiddish expression *mazel tov* is derived from Hebrew words that mean, in effect, that one lives under favorable stars; *mazel* originally meant the influences on our lives that rained down from the zodiac.

The Jewish historian Josephus in *The Jewish War* refers to astrological signs when describing the destruction of the Temple:

> Thus were the miserable people persuaded by these deceivers, and such as belied God himself; while they did not attend nor give credit to the signs that were so evident, and did so plainly foretell their future desolation, but, like men infatuated, without either eyes to see or minds to consider, did not regard the denunciations that God made to them. Thus there was a star resembling a sword, which stood over the city, and a comet, that continued a whole year. (Book IV, Ch 5, Sec. 3; as translated by William Whiston in *The Works of Josephus*)

(Notice how, as we mentioned above, the comet was understood to be a sign of doom.)

There is a Jewish parallel to the Matthew story in a Midrash on the birth of Abraham, which describes how his birth was foretold by astrologers as a threat to the king of Babylon, with the infant Abram being hidden for three years from the soldiers of the king. Scholars can argue how much this story and the Matthew story influenced each other; for our purposes, it's enough to note its existence as evidence of the complex attitude of Judaism toward astrology.

Of course, today astrology is in bad odor among scientists for a much simpler reason. It doesn't work.

PAUL: What do you mean when you say, "It doesn't work"? Though astrology doesn't accurately forecast future events, it seems to "work" on other levels for some people: they use it to bring some sort of meaning and order into their lives. Astrology seems to meet psychological and emotional needs for security, for meaning, and so forth. So even though astrology isn't scientific, by modern standards, it still seems to "work" for some people.

GUY: What I mean by "not working" is that ancient astrology was based on certain fundamental assumptions about the universe that we know today just aren't true.

We talked about that the first day: the cosmology of ancient Greece had Earth encircled by a few planets, each in its own sphere, all inside the sphere of the stars. A whole host of "planetary intelligences" were thought to inhabit those spheres, with *daemons* (Greek) or *genii* (Latin) communicating "influences" from sphere to sphere. One such *genius* was the personification of Sleep, another that of Love; other *genii* carried special gifts such as talents in music or mathematics—the origin of our usage of "genius" today. So, the theory went, if you had a talent for music (or sleeping, I guess), it meant that you were born at a time when one of those geniuses was overhead.

But we know today that none of that cosmology is correct; none of those hypothesized spheres actually exist. We have a completely different way of describing the orbits of planets today, and a completely different idea of what those planets are.

Furthermore, if you try to test astrology scientifically, such as asking it to predict phenomena and future events and then checking to see whether those predictions are borne out, the result inevitably is that astrology gets falsified, disproved.

PAUL: For the famous philosopher Karl Popper, astrology was a standard example of a "pseudoscience"—along with Marxism. According to Popper, both disciplines make "predictions" that are so vague and general that they can't ever really be shown to be wrong. That was Popper's criterion of what makes science scientific: a science must make predictions that can be shown to be wrong. But still, as recently as Galileo and Kepler's time, astronomers were expected to cast horoscopes. That's how they made their living. It was only later that astrology got cleanly separated from astronomy.

GUY: So, I come back to my question: Why did Matthew (or God) choose astrologers, of all people, to find Jesus? To me, that's a puzzle. It doesn't have the simple logic of a story that was just made up. It's so strange that it makes me suspect that Matthew is describing something that actually happened.

PAUL: Nowadays we think of astrology as being mainly concerned with trying to predict the future. But back before the rise of modern science, that wasn't the main concern of astrology. In those days, astrology was valued as a system of meaning, in which prediction played a *part*. The point of prediction was not so much to tell you the future as to help you make sense of the *present* meaning of your life; prediction was meant to help you see who you are in your entirety—past, present, and future all together as part of your identity.

GUY: I'm reminded of the feeling you'd get watching that TV show, *The Young Indiana Jones Chronicles,* knowing that the main character would grow up to become Harrison Ford.

PAUL: If that's what astrology was about, then it makes sense to me that astrologers would be the ones to find Christ. They would be expected to be concerned with recognizing the whole person, in his fullness. They would be expected to be sensitive to the identity and importance of Christ.

GUY: Nothing we've said so far proves by any means that the "pious story" explanation isn't the real one. Just because we can't understand the way the story was put together doesn't prove that there weren't perfectly good reasons for it. Just because I don't understand something doesn't prove it isn't so. We must also accept the possibility that, rather than a factual account of an actual astronomical event, the story in Matthew was intended as a parable with a message that people in his day could hear and exclaim, "Aha!" with comprehension, in a way that perhaps we don't fully understand today.

PAUL: So maybe the Star was a real astronomical event that occurred at the time Jesus was born, or maybe the whole story was invented but with the intention to express important truths about Jesus Christ.

A MIRACLE?

PAUL: Guy, did you ever have an "aha!" moment with these telescopes?

GUY: I remember one event quite clearly. The summer of 1994, my second year here, everybody in the astronomy world was all excited about Comet Shoemaker-Levy 9. The comet itself wasn't much to see; in fact, it had already passed close by Jupiter and gotten itself ripped into a couple dozen pieces. But those pieces were about to do something spectacular.

The guys who do orbital mechanics had calculated the paths of the train of comet bits and determined that they were all about to hit Jupiter itself, *pop-pop-pop-pop*, over a period of about a week. And here in Rome, we were perfectly poised to observe the first strike!

No one had ever seen a comet hit a planet before. We had no idea what to expect. And, even worse, the calculations said that it was going to hit the far side of Jupiter, away from where we could see it. So what could we observe? One idea was to look at one of Jupiter's moons, one that was in a position to reflect whatever flash of light might occur during the impact, so we could estimate when the first piece of comet actually hit and how powerful a flash it made. I set up a camera on this telescope and aimed it at Jupiter's moon, Io, hoping to see such an event.

Meanwhile, two amateurs—pretty good amateurs, as you often find in amateur astronomy—had set themselves up at another one of the observatory's telescopes, the big refractor in the dome next to us here atop the Pope's vacation home. They had (what was then) a new gizmo called a CCD camera, attached to a personal computer, to take images of Jupiter itself. I didn't expect they'd be able to see anything; I mean, the impact was on the wrong side from where they were looking. But if they wanted to use our other telescope and take a look, why not? No one else was using it.

When the impact was supposed to have occurred, our camera detected nothing. No flashes of light. Turns out we were looking in the wrong filter, the wrong color, to see the flash well, but we didn't know that at the time. Disappointed, I wandered over to the other dome to see how my friends were doing. Just as I got there, I heard them shout. The image of Jupiter from their CCD camera was displayed on their computer screen, and now, half an hour after the event, the part of Jupiter that had been hit by the comet had rotated into view. And there, on the cloud tops of Jupiter, was an enormous black spot where the comet had hit!

We had no idea that the comet would do that to Jupiter. No one had ever seen anything like that before. But since we were in the right place to see the first impact, we were among the first people in the world to see the black spot that Comet Shoemaker-Levy 9 made when it hit Jupiter.

And, incidentally, we were also able to see how well those CCD cameras worked. Everybody uses them now.

Anyway, that was a real "aha!" moment for me.

PAUL: I want to come back to that point—I think that "aha!" experiences have something to do with miracles. But first I need to ask you something. I thought these telescopes here atop the Pope's summer residence were no longer suitable for research-level astronomical observation. They're lovely instruments, but I thought that the light and smog from Rome now make it impossible to do research-level astronomy from here. That's why the Vatican Observatory has a modern telescope outside Tucson, Arizona, right? But now you're telling me that some honest-to-Pete observational astronomy got done with these telescopes right here, not all that long ago?

GUY: We can still do certain kinds of research-level astronomical observation from here at Castel Gandolfo. About ten years ago, there was going to be an occultation of a faint star by Pluto: the sort of event we talked about back in Antarctica on Day 2, where you measure how big Pluto is by timing how long it takes for a star to blink on and off while Pluto passes in front of it. A buddy of mine from Arizona had shown up here with a high-speed electronic camera to record the event.

But, of course—wouldn't you know it?—at sunset the sky was completely clouded over. So the two of us shrugged, chalked it up as a bad night, and went for a pizza at Sor Capanna—the restaurant you mentioned earlier, when we first came up here. We enjoyed a bottle of wine with the pizza to ease our troubles. Maybe two bottles. When we walked out of the restaurant a few hours later, a little unsteady on our feet—wouldn't you know it?—the sky had cleared up completely! It was a perfect night! And there was still half an hour before the occultation was supposed to occur. So we rushed up to the telescope dome here and got to work.

It felt a little scary, driving a big telescope with a bit too much wine inside me. I don't know which was the bigger miracle—the change in the weather, or the fact that we were able to operate the telescope flawlessly and get our data that night.

PAUL: I'm not sure that successfully driving a telescope after consuming a pizza and two bottles of wine would qualify as a miracle.

But what makes a miracle a miracle? What do we mean if we call the Star of Bethlehem a miracle? Do we mean that it was zooming about the sky like some sort of UFO, guiding three wise men to a stable in Bethlehem?

GUY: There's nothing in Matthew that says that the star darted about in the sky or, for that matter, that the number of wise men was three. And by the time of the Magi's arrival, you would hope that the Holy Family had been able to move out of the stable!

PAUL: Fair enough. We need to be careful about mixing up what's in the Gospel with what comes from later, much-loved stories.

GUY: More to the point, making the star into that kind of "miracle" is theologically suspect. Sure, an all-powerful God could have created such a UFO. An all-powerful God can do pretty much anything He wants. Such a God could have sent Jesus into the world fully grown and powerful, unmistakably a deity, dressed up in a man-suit like some Eastern avatar—the expected image of the Messiah, one that no one could have mistaken.

Instead, God exercises a kind of supernatural restraint: Jesus comes into the world as an infant, born like any other human and subject to the very laws that God used to form this universe.

PAUL: The idea of God being "all-powerful" entered Christianity from Greek philosophy. The God of the Hebrew scriptures is very powerful, powerful enough. But the Hebrew scriptures don't make any sort of big deal about God being *all*-powerful.

GUY: If God had not shown supernatural restraint, science would not be possible. Science is possible only if natural phenomena have a character that is lawlike and consistent. A God who acted willy-nilly in the world, "interfering" in natural processes right and left, would leave no room for science. It is God's restraint that permits the natural world to run its course in a way that is consistent and lawlike. And so it's God's restraint that makes it possible for us to do science.

PAUL: What you call "supernatural restraint" can also be understood in terms of the traditional distinction between primary and secondary causality. In one sense, God leaves the world alone: He allows it to run its course autonomously, according to its own laws. But in another sense, God is intimately involved in the world: He holds the world in being, and He sustains the laws of nature. God is the "primary cause" of the world in that He gives it order and sustains it in being. But, as a result, the world has its own consistent order and causal laws—its own "secondary causality."

With the rise of the new science in the seventeenth century, which saw the world as being a kind of mechanism, there was great progress in understanding secondary causality: many laws of nature were discovered and understood in a new and deeper way. But the role of God as primary cause faded into the background, as far as science was concerned. Science came to see God as being a kind of "watchmaker"—as one who creates the mechanism of the world and winds it up but then lets it run on its own, without interfering in any way. The notion of a God who sustains the world in being and underwrites the laws of nature wasn't something the new science could grapple with—it just didn't fit.

But this meant that miracles came to be understood in a new way—the way we tend to understand them to this day, in fact. If you see nature as being a kind of inexorable machine or mechanism, one whose motions are utterly determined and predictable in accordance with rigid mathematical physical laws, then miracles come to be seen as violations of those laws; the only way God can intervene, the only way God can act in the world, is by violating the laws of physics. That's how miracles came to be understood in the sixteenth and seventeenth centuries, as the world itself came to be understood as being like a machine or mechanism.

GUY: But wait a minute . . . if God is all-powerful, all-knowing, and all-good, why would He ever create a world in which He needed to suspend the laws? Shouldn't God be, not just a "watchmaker," but a *perfect* "watchmaker"? Shouldn't such a God create a world in which the laws of physics are so perfect to begin with that He'll never need to meddle or tinker later on? In this way of thinking, every time God intervenes miraculously in the world, it's a sign of weakness and failure on His part—a sign that He didn't do the job of Creation right to begin with!

PAUL: Exactly. That's the problem with our modern way of understanding miracles. It's this way of thinking that led many people in the seventeenth and eighteenth centuries to become *Deists*. They saw it as God's job to create the world and then leave it alone. And it's this way of thinking that has led people in our own day to think that they must choose between science and religion, between reason and faith: between a rational science that leaves no room for God to act in the world and a faith that transcends and violates reason but leaves room for God to act in the world.

GUY: It's a central pillar of Christianity that God does somehow act in human history. The Christian God plays a big role in human history: most obviously in the life, death, and Resurrection of Jesus Christ, but in other ways, too.

If you think that a miracle must involve some sort of violation of the laws of physics, that puts God in an impossible position. God would only be able to act in the world by violating the laws of physics that He Himself created and sustains. That is a game that is rigged to make it impossible to conceive of any rational way for God to act in the world. But that is the way of thinking that is presumed by fundamentalists, both scientific and religious. That way of thinking leads religious fundamentalists to reject modern science as contrary to faith. And it leads scientific fundamentalists to reject faith as contrary to science.

PAUL: But prior to the scientific revolution, miracles weren't understood as being violations of the laws of physics. That's certainly not how miracles were understood in Scripture.

People were talking about miracles long before the concept of a "law of physics" ever existed. What was it that made a miracle a miracle, in those days? The reason an event was considered miraculous was not that it involved a violation of physical law, but because it was an important sign that revealed something about God or the Kingdom of God. A miracle might or might not happen to involve something out of the ordinary course of nature. But that was not really the point. The important thing about a miracle—what made it a miracle—was that it was a *sign that revealed God and God's Kingdom.*

Something can be a "sign that reveals" without involving any violations of the laws of physics. The rainbow that Noah saw after the flood was an important and miraculous sign of a new covenant with God—but it did not involve any violations of the laws of physics.

Look, suppose a husband packs his wife's lunch every day, and he always includes a penny in her lunch bag. To anyone but his wife, the presence of the penny is meaningless. But it so happens that on their very first date, the two of them threw a penny into a fountain and wished for a long life together. So, for the couple, the penny is a sign and symbol of their love—the penny in the lunch bag reveals and reinforces their love, making it all the more present. The penny is a remarkable sign and symbol that is also efficacious—it has a real effect on the love of the couple. But physically the penny is utterly unremarkable—it involves no violations of the laws of physics.

That is the way to think about divine miracles: whether or not they happen to involve a violation of the laws of physics isn't the point. What makes a miracle a miracle is that it reveals God and God's Kingdom in a remarkable way.

GUY: I do believe God acts in the universe. A "divine watchmaker" who winds things up and then walks away does not describe the God of love I experience in prayer, the God I see acting in my life every day.

But one thing I've noticed about God's action in the world . . . He always seems to give you "plausible deniability." If you really don't want to accept that He is present, He doesn't force you to. You always have the option to not believe, to come up with some alternate explanation for anything He does.

PAUL: I am not claiming that God cannot or does not violate the laws of physics, as we understand them. The Resurrection of Christ, the central sign and miracle of Christian salvation history, would seem to be something difficult to reconcile with our usual understanding of physical laws. Still, I think we're missing the boat on miracles if our primary focus is on whether or not God has violated some law of physics. If you are calling in scientists to help you figure out whether or not a miracle occurred, you're missing out on what it is that a miracle is all about.

GUY: So how should we respond when someone claims that a miracle has occurred?

The stereotypical "scientific" response to observing some unusual event is to try to account for everything we see in terms of the already-known laws of science. If the known laws of science can't explain some strange or unusual event, then rather than claiming a miracle occurred, we're more likely to doubt that the event itself actually occurred. If it can't be explained in terms of the science we know, then probably it didn't happen.

PAUL: This is the basic conservatism that is built into modern science: the strong bias is to accept as actual or possible only things that are like things that have been seen before, and only things that are in accord with accepted physical laws. So the bias is to be skeptical about events that are strange or unusual.

GUY: But sometimes the world can surprise us. For example, Thomas Jefferson insisted that the reports in 1803 of a meteorite fall in France must have been the fabrication of credulous Frenchmen. Since he'd never seen a meteorite fall himself, he found it hard to believe that such a thing could ever happen. Given that I work in a laboratory with a collection of a thousand such meteorites, many seen and, indeed, filmed while they fell, we know now that bits of outer space do impinge upon our world, fall through our skies, and touch our Earth. It's rare, but it happens.

And we recognize that a little humility is always in order before asserting what can, and cannot, occur in nature . . . if nothing else, our formulations of the laws of science are never final and undoubtedly will look very different in a thousand years' time.

PAUL: Prior to the rise of modern science, there was less skepticism about strange and unusual events. In those days, instead of thinking of the world as being like a big machine whose parts fit together and moved with mathematical precision, people thought of the world as being like a living organism, whose parts moved like the parts of a plant or animal. Machines always act the same way, unless they break. But that's not the case with living things. Though they usually act the same way, for the most part, sometimes they can act a bit differently.

Suppose we know, from long experience, that Uncle Charlie loves dogs and hates cats. It will seem remarkable to us if one day he shows up holding a cat, petting it and talking sweetly to it. That would be contrary to Uncle Charlie's usual character and way of behaving. It might be a sign that something unusual or important is happening in his life. But we would never think that his behavior is a violation of some physical law.

GUY: Maybe he's got a new, cat-loving girlfriend. Maybe he's had a stroke.

PAUL: Prior to the rise of modern science, things in the world were understood to have characteristic ways of behaving, just as various kinds of people have characteristic personalities. So the world was understood to have a sort of built-in variability: things usually behave in certain characteristic ways, but sometimes they behave differently—like Uncle Charlie. The point is that when something really unusual happened, it was not automatically taken to be a violation of physical law. It was just taken to be an unusual event—a "wonder" or a "prodigy."

GUY: So something was a wonder or prodigy if it was highly unusual, out of the norm. And something was miraculous if it happened to reveal something important about God. But there were wonders and prodigies that were not considered miraculous: they weren't considered to reveal anything about God. And there were miracles that weren't wonders or prodigies—like Noah's rainbow.

A MYSTERY?

PAUL: The sun has set now, and it's gotten pretty dark here in the dome. I have to admit, it's rather pleasant—if chilly—here in the dark. I can hear the sounds of the town below us, coming through the slit opening of the dome, but it's like they're all at a distance; we're cut off from them. The way the noises echo off the inside of our wooden dome makes it all seem quite mysterious. It's funny, you always think of science as being rational and orderly, and you forget that there's a real pleasure to it, too, being alone in a place with an atmosphere like this. I like it. I like that mysterious feeling. It's a fun kind of mystery.

GUY: So, what do you think? Do we have to write the Star of Bethlehem off as a mystery? We weren't there, and I think we'll never know for sure if Matthew was talking about something unusual in the sky that actually occurred at that time. So I think that the Star of Bethlehem must remain a mystery.

PAUL: A mystery in what sense?

For scientists, a mystery is a problem to be solved—it's a gap in our knowledge that needs to be filled.

For example, Isaac Newton solved a mystery that was created by Johannes Kepler's discovery of the elliptical form of the orbits of the planets. Long before Kepler, everyone considered it obvious that planetary motions should be circular, since that was the most perfect form of motion. Kepler's discovery that planetary orbits were, in fact, elliptical created a problem, a mystery to be solved: why are the planets' orbits ellipses and not circles? The mystery was solved by Newton: elliptical orbits are just what is predicted when you combine Newton's three Laws of Motion with his inverse-square force Law of Universal Gravitation. Kepler's new discovery had created a mystery; Newton's new theories solved it.

Of course, Newton's discovery created a new mystery: how does the force of gravity act at a distance?

GUY: Before Newton proposed his Laws of Motion, that question wouldn't have been asked; the very words would have been meaningless.

PAUL: That's a feature of scientific progress: there's a mystery to be resolved that captures everyone's interest. But as soon as the problem gets resolved, the sense of mystery evaporates—the answer seems obvious. In fact, what had seemed mysterious becomes a taken-for-granted part of scientists' regular "toolbox."

Since that is how science progresses, it is how science tends to be taught. Science teachers pose "mysteries" to their students, in the form of problems to be solved or experiments to be performed. The teachers already know the answers, of course. From their perspective, there is nothing particularly mysterious about them (apart from the mysterious ability of students to confuse the issue). The teachers try to set things up so that their students will have the experience of solving mysteries and discovering laws of science on their own. Of course, older students are given problems and experiments that are more complex. And with their best and most advanced students, the teachers share problems that no one has yet solved—problems that really are mysterious. Solving those problems earns you a PhD.

GUY: Scientists need problems to work on. We need mysteries. The goal of a scientist is to solve mysteries—to eliminate them, to make them disappear. When we succeed, what once seemed mysterious and interesting becomes obvious and pedestrian. *And then new mysteries emerge, and the cycle of discovery and explanation continues.* That's the function of mystery in the world of science.

PAUL: That's what mystery means in science—and in detective novels. But in the wider domain of human life and interaction, as well as in the domain of faith and religion, the word *mystery* is used in a different way.

A religious *mystery* is not a problem to be solved, or a gap in our knowledge that needs to be filled. Believers do not seek to eliminate mysteries or to make them disappear. Instead, they seek instead to *deepen* the mysteries, to *dwell in* the mysteries, to *savor* the mysteries. Believers use the word *mystery* to indicate a point of connection between the human and the transcendent or divine. Mystery is where the human capacity for understanding gets swamped—not because we haven't yet figured things out, not because there is a problem that remains to be solved, but because the proper response to God and to love is not to understand, but rather to treasure and to ponder.

The religious approach to mystery is encapsulated in a little passage in the second chapter of Luke's Gospel. After describing the signs and wonders associated with the birth of Jesus, Luke says that Mary "treasured these things and pondered them in her heart."

Mary recognized that it was important to *treasure* and *ponder* what had happened, even if she could not understand it all. This became the pattern for Mary's life, as memorialized in the Catholic tradition of the mysteries of the rosary. The mysteries of the rosary represent deep, important events in Mary's life that she treasured and pondered in her heart. Since the events of Mary's life happen also to have been important events in the history of Christian salvation, religious believers treasure and ponder those same events today when they pray the rosary.

For scientists, the goal is to solve the mysteries that confront us—to make them disappear, so that new mysteries will emerge and the game can go on. But that is not at all the attitude of Mary and other believers who pray the mysteries of the rosary. For them, the goal is to *dwell in* those mysteries, to enter more deeply into them, to treasure and ponder them. From the perspective of believers, these mysteries mark points of contact with the transcendent or divine.

Consider love. Love is mysterious in both senses, scientific and religious. From the perspective of science, love is a problem to be solved: the task is to describe the phenomenon of love and to understand its mechanisms, the associated neurophysical brain states, biochemical hormones, social patterns of behavior, etc. Since these features and mechanisms of love have not yet been fully understood and explained by science, love remains scientifically mysterious: it is a problem to be solved.

But even if one day we arrive at an adequate scientific account of the phenomenon of love, love will remain mysterious on another level. The experience of being in love will still swamp human understanding—it will still be a point of encounter with the transcendent. The appropriate human response to love is not to try to solve the problem or to eliminate the mystery, but to enter more deeply into the mystery—to love more deeply, and to treasure that love and to ponder it in your heart.

Is the Star of Bethlehem a mystery? From the perspective of science, no. The question of whether there really was something remarkable in the heavens at the time of the birth of Jesus and, if so, what was it, can be attacked with all the tools of science and other human disciplines. These are questions that have factual answers—questions that could be solved. In theory, if we had a time machine and a video camera, we could travel back to the appropriate year and interview the Magi—or Matthew—and find out definitely what it was they were talking about. We could observe the light of the Star itself and determine where it came from. But as we've seen, there are multiple possible scientific explanations for a phenomenon like the Star of Bethlehem, if it occurred. So there is no fundamental scientific mystery. There is just the unresolved historical question about whether a particular event did or did not actually occur.

But from a wider human perspective, the Star remains a mystery. For the Magi in Matthew's story, the Star was a point of encounter with the divine. The Magi's reported encounter with mystery, by being recorded and shared, becomes at secondhand our own encounter with mystery. This is similar to what happens in the case of the mysteries of the rosary: Mary's encounter with mystery becomes our encounter with mystery.

A GUIDING STAR

PAUL: That sounds like quite a crowd out in the square below us. We can hear the racket even up here in the dome. There's even a brass band playing! I wonder what's going on.

GUY: It might be folks coming out of the church in the square, Saint Thomas Villanova. There's a crèche scene set up in the church hall that's quite popular; I think tonight's the night it was opened to the congregation to view. It attracts a good crowd every year. And then, of course, after you've seen the crèche, you can go into the square, have an espresso, and chat with your neighbors. I think that's what's going on.

PAUL: You think. But you don't know. I suppose we could actually go down and look for ourselves.

GUY: Now who's being the scientist?

But back to the Magi: you know, the Magi in Matthew's Gospel are an inspiration and guide for all the Jesuits who have worked at the Vatican Observatory, from the ones who built these domes eighty years ago, to the ones working at our modern telescope in Tucson, Arizona, today.

As *scientists*, we Jesuits seek to understand the stars and the heavens, like the Magi. Our goal is to make some scientific mysteries go away and discover some new ones. Look around at this telescope dome where we're standing. Just think about how the astronomy we do today is different from the astronomical questions that were hot back when these telescopes were built, eighty years ago. Our questions are different now because of the success those Jesuits had with these old telescopes back then—through their work they resolved some old scientific mysteries and engendered some new ones.

For example, back when these telescopes were built, people measured the spectra of stars by putting a prism in the path of the starlight and measuring the intensity of the resulting rainbow as recorded on a piece of photographic film. Nowadays we use grating spectrometers and CCD chips—unlike prisms and film, they give a "linear response" to the data that makes it much easier to compare against data observed with other telescopes. No journal today would publish new data taken on photographic plates. (In fact, nobody even makes those plates anymore.)

The questions we ask today are shaped by the technology available now that we didn't have then, or by the technology we hope to have in the future but don't have yet. As scientists, we stand on the shoulders of giants, even as we strive to surpass them. That's how science works: you respect and venerate your predecessors, but you also seek to leave them behind—you seek to render them anachronistic. To the extent that we see the Magi as scientists, we've left them far behind—they are no longer relevant to us.

But as *men of faith* we Jesuits are alert to points of encounter with the divine, where we are invited by God not to resolve the mystery but rather to enter more deeply into the mystery. As the slogan on the side of this dome says, we seek not only to understand the world but also to come and adore God the Creator. In that sense, we do not stand on the shoulders of the Magi, and we have not left them behind. We stand side by side with them: they are our companions in adoration.

Two thousand years ago, the Magi, following their calculations, were led to a king very different from what they were expecting. Interestingly, the story tells us that it was outsiders, foreign intellectuals, who found the Messiah before all the experts of the Temple did. This could be one more example of how God can use even our foolishness, even our astronomy, to lead us to Him. That is why I see the Magi as models for us Jesuits here at the observatory: doing science, they are surprised to encounter God. And their thirst to find God motivates them to do science.

PAUL: The challenge for us is to confront the world both as scientists and as seekers: to be open both to understanding the world on its own terms and to seeing the world alight with divine mystery.

GUY: "Scientists and seekers?" Cosmic, man! I don't know, that sounds a little too New Age to me. But I know what you mean.

Sometimes I work with passion on solving the puzzles of astronomy; sometimes I just like to look at the stars.

How about, "ponderers and puzzlers"? When I am doing my puzzle thing, I might ask: What could the Star of Bethlehem have been, if it really existed—what known natural effects are compatible with the reported observed phenomenon? How likely is it that some such phenomenon actually existed? And if it's unlikely that such a phenomenon actually existed, can we explain why it was reported? Those are questions of fact and causality, which I can ask and pursue as a puzzler. It turns out that at the end of the day astronomy does not give us a unique, singular answer, but that's pretty typical, actually. There are often several more or less likely possible explanations for a given phenomenon.

But as a ponderer and stargazer, I might ask: what was it like to see Christ as the Magi did, from their perspective?

The most astonishing part of the story of the Magi, from my perspective, is not that they would predict the birth of a king from the positions of the planets; that kind of calculation is merely mechanical—any fortune-teller could have done that. Nor is it that they'd pull up roots and travel afar to find out if they were right; astronomers do that all the time, when they head for distant telescopes. Instead, it's that they would be able and willing to recognize the child they found to be the king they were seeking.

It's easy enough today to follow the crowd, like the one out in the square tonight, to the manger scene and mouth all the right words about peace on Earth. But two thousand years ago, the shepherds and the Magi had only their hunches to follow. Were those really angels singing? Does that star mean something? How did they know?

The wise men and the fishermen, the prostitutes in Judea and the wealthy women in Macedonia . . . what stirred them to jump to this new, strange Gospel of confessing and forgiving sin, to follow a baby born in a stable whose teachings got him hung upon a cross?

PAUL: During the season of Advent, leading up to Christmas, we Christians are taught to pray that we might know and recognize our deepest desires and longings. Their fulfillment isn't always found where we'd expect. The best presents are not necessarily the ones under the tree.

GUY: During Christmas, we celebrate the arrival on Earth of a heavenly visitor just as real as any of my meteorites but even more shattering. When things like that arrive, out of the blue, our own narrow worldview is disrupted, and we are forced to recognize a universe bigger than our mundane lives. Both natural science and humble religion inform us: remarkable things do occur. And whether or not there was a visible Star, there certainly was a visible Savior. That alone is miracle enough.

That miracle is echoed in all the other unexpected signs of love, large and small, that fill our lives. We're standing inside one such miracle right here, perched on a hilltop as high as that comet across the lake, and just as visible, year-round. Think of it . . . at the height of the Depression, in the early 1930s, a Pope agreed to spend the money it took to build these two telescopes, with fine Zeiss optics, state-of-the-art back then, on the roof of his summer home . . . where we're standing right now.

Every miracle is like cracking open the slit of our telescope dome. We are surrounded by mundane concrete and metal and wood, but through the slit we suddenly are able to peer out at a universe full of stars.

Speaking of which . . . want to take a look through one of these telescopes? The Orion Nebula is rising.

PAUL: Sure—I never get tired of looking at Orion. As long as we don't take too long; it's cold with that dome open!

GUY: Just one peek. Then we can head out to Sor Capanna and get a pizza, hot from their wood-burning oven!

DAY 5: What's Going to Happen When the World Ends?

SETTING: DINING AT THE END OF THE UNIVERSE

ANTIPASTO

PAUL: Hello, waiter? We'll start with the bruschetta with bacon and balsamic vinegar, and also some with prosciutto and mozzarella.

GUY: Funny, how we always seem to end up ordering too much. After the pizza last night, now this. Oh, my aching arteries!

PAUL: Our aching arteries are reminders of our mortality, Guy. We aren't eighteen anymore. We're both well beyond thirty-five, the midpoint of the three score and ten years traditionally allotted for human life.

Wow, there's a sobering thought. Pass the wine, please.

GUY: And we don't know . . . we may be practically at the end of our lives, or we may have "miles to go before we sleep."

You're right, a sobering thought. I'll have that wine back over here.

PAUL: Just as we can wonder how much longer each of us will be around, we can also wonder how much longer the universe itself will be around. Is the universe young, old, or middle-aged? Could it come to an end tomorrow, or will it still be here tens of billions of years from now—and what will it be like by then?

GUY: By the way, this bruschetta is fantastic. Good choice.

PAUL: If we keep eating at this rate, we'll really be putting on some serious weight. Maybe we should worry about that, too—but not tonight!

GUY: But it's not just our waistlines that are expanding. The whole universe is expanding.

PAUL: Some people are worried about that. Do you remember the scene from Woody Allen's Oscar-winning movie *Annie Hall,* when the main character, Alvy, flashes back to a childhood visit with his mother to the family physician? His mother is worried about why little Alvy is so depressed. "It's because the universe is expanding," he explains. "And someday it will break apart, and that'll be the end of everything!"

The doctor reassures him that that won't happen for billions of years and tells him that in the meantime he should distract himself from his worry and enjoy life while it lasts. His mother is more direct: "You're here in Brooklyn! And Brooklyn is not expanding!"

GUY: I assume Woody Allen was just making fun of that kind of angst . . .

PAUL: No, I think he was serious. In an interview in *Esquire*, in September 2013, Woody said that our presence on Earth is an accident, and everything we value in the universe will someday be gone, "whether it's Shakespeare, Beethoven, da Vinci, or whatever," along with the Earth and the sun.

So Woody thinks science has shown that the future universe will be inhospitable to human life, and also that the presence of human life in the universe is a mere accident. And he concludes that there's really no point to anything. There's no point in asking "Big Questions"; since nothing will last, nothing is of ultimate value. The best we can do is distract ourselves—with work, with love, or whatever—and enjoy life while it lasts.

GUY: But it seems clear that "Big Questions" do matter a lot to Woody. He's spent a lot of time thinking about them, asking about them; they come up in his movies all the time.

I wonder about being fixated on what happens at the end of it all. Isn't that just treating the universe like a big murder-mystery novel, where the goal is to find out, on the last page, whodunit?

PAUL: Then there are those who think it is pointless or presumptuous for us humans to ask "Big Questions." This is a theme that runs all through Douglas Adams's comic work *The Hitchhiker's Guide to the Galaxy*. In the *Hitchhiker's Guide*, humans are constantly reminded that their existence is so puny and insignificant, relative to the size and scale of the universe, that it is foolish for them to think themselves capable of asking questions that are really big. Even if you don't agree, it's a point of view that's worth mulling over and chewing on.

GUY: So that explains where we are! Tonight we're eating at Milliways, the fictional Restaurant at the End of the Universe, from *The Hitchhiker's Guide to the Galaxy*.

PAUL: We've come to this restaurant by traveling forward through time, billions and billions of years into the future, to a point just a couple of hours prior to the end of the universe. While enjoying a delicious, five-star meal, we'll get to watch through the windows as the universe comes to an end all around us. (The restaurant is somehow protected from destruction by a time-space force field. Don't ask.) After the meal, we'll travel back through time to our own era, billions of years in the past.

GUY: Just my luck. It's my one chance to pig out like there's no to-morrow, and the food is all imaginary. By the way, can I have an-other bit of that bacon-and-balsamic bruschetta?

How did we ever manage to get a table here, by the way? From what I hear, this place tends to get booked solid, eons in advance. And it's not as though fictional restaurants are all that easy to find.

PAUL: I pulled strings with some of your old science-fiction writer friends—they managed to get us a table here tonight.

You would think that dinner at Milliways would be a once-in-a-lifetime, awe-inspiring event. But that's not how it gets portrayed by Douglas Adams. In *The Hitchhiker's Guide*, the End of the Universe (visible through the restaurant windows) is reduced to the status of being just part of the floor show—like a pratfall punctuated with a drumroll. Douglas Adams challenges our presumption by portray-ing the End of the Universe as something tawdry and insignificant, about as noteworthy as the entertainment at a second-rate Catskills dinner theater.

GUY: Hey, my great-uncles were vaudeville performers; a hundred years ago they made a good living in the Catskills!

But look, we're getting ahead of ourselves. Before we can decide whether or not our "Big Questions" really are significant—before we can decide whether the ultimate fate of the universe really matters—I'd like to get back to what Woody Allen and Douglas Adams think about the end of the universe.

PAUL: OK, let's go back to the science. For starters, Woody Allen thinks that science has shown that the universe is expanding, and that in the far future it will become inhospitable to human life. Did he get that right? What can science tell us now about the far-future state of the universe? How reliable is our science on that question?

GUY: That's not so easy to say. The trouble is this: we can have scientific theories about the beginnings of things, the origin and evolution of the universe, as we discussed on Day 1 of our conversations back in Chicago at the Art Institute. Looking deep into space, we are seeing photons of light that left their stars billions of years ago, and so we can see what things looked like back then.

Closer to home, theories about the origins of our own planet and solar system are relatively easy to test, because we have bits and pieces preserved from the past that we can compare against our theories—rocks from the moon and the asteroid belt, my precious meteorites. We have samples in our lab that have lain essentially untouched since they were formed billions of years ago. They can tell us about the conditions that prevailed when they and the planets were made.

We have data from the past. So we can talk with some confidence about what happened in the past. But we have no samples, no data, from the future!

PAUL: But many of the equations and laws of modern physics are time-symmetric. So as far as much of physics is concerned, retrodiction and prediction—modeling what happened in the past and figuring out what will happen in the future—are on the same level. I mean, if you know what the locations and motions of the planets are right now, you can use Newton's laws to predict where they'll be in the future, ten or a thousand or a million years from now. And you can just as easily work backward to retrodict where they were in the past, ten or a thousand or a million years ago. As far as the math is concerned, there's no difference between past and future, between prediction and retrodiction.

GUY: While that may be so, not *all* the laws of physics are time-symmetric in the way you just described. But more to the point, we don't use data from the past merely to set the "boundary conditions" to plug into our neat little physics theories. We use that data to fill in and cover over, or at any rate keep on track, all the inadequacies of our terrible, horrible, no good, very bad (but the best we have, in spite of that) theories. We understand so little, still, of how the universe works that we can only connect the dots when they are close together, so to speak. Data from the past give us lots of dots to connect. No data from the future? We're flying blind.

PAUL: Just the point I was getting to. Can we really be sure that the laws of physics, as we know them at this point, are sufficiently correct for us to be able to rely on them for predicting the far-future state of the universe?

And even if the laws of physics as we know them at this point are correct, can we be sure that those laws won't change in the future? All the evidence that we have to support the laws of physics comes from the past. To make use of those laws to predict the future, we have to assume that things will keep working the same way in the future as in the past. And that is an assumption—it is not something we can prove. Thank you, David Hume.

GUY: David Hume . . . philosopher, right? I think I had to read something by him a long time ago.

PAUL: Eighteenth-century Scottish empiricist.

So extrapolating accurately to the far future is difficult for several reasons. As Hume points out, we can't know with certainty that the laws will stay the same in future. If you argue that, based on past experience, the laws of physics will remain the same in the future, you've just presupposed the very thing that you're trying to prove. Still, we have no choice but to assume the stability of the laws of nature, if we're going to be able to do any science at all.

When it comes to predicting the far future, even small errors in the accuracy of the laws of physics we now take to be true and even small errors in the accuracy of our knowledge of the current state of the universe could, over the long term, compound exponentially and make our predictions wildly wrong. You need accurate laws and accurate knowledge of the state of the system at a particular point in time, and the longer you let the system evolve, the more your errors will be compounded. If the laws of physics are nonlinear, or liable to go chaotic, then even a tiny uncertainty can suddenly grow rapidly and utterly unpredictably.

GUY: That said, little Alvy is correct to this extent: right now, at least, we can see that the universe is expanding. Granted, theories about the future of the universe are impossible to test, and the best we can do is assume that whatever is happening now will continue to happen the same way into the future. But if space continues to expand forever, as most versions of the Big Bang theory predict, then eventually the various galaxy clusters will be so far apart that they will lose all possibility of contact with each other—light shining from one will never be able to reach the other.

Mind you, experience tells us that such extrapolations are always very uncertain. That prediction simply represents the least-unlikely outcome. Other theories out there suggest completely different outcomes for everything in the far, far future . . . the laws of gravity could change with time, such that the Big Bang could turn around and become a Big Crunch, or boundaries between different dimensional realms of the multiverse, called branes, could sweep past us and wipe out everything in an instant, without notice!

But it turns out that, even if the standard Big Bang theory with its eternal expansion is true, little Alvy's pessimism is misplaced.

The expansion of the universe isn't the problem; our own galaxy will still hold together. His mom is correct: Brooklyn isn't expanding. Distant clusters of galaxies may be drifting away, but who needs them? They are invisible to the naked eye; we didn't even know they were there until the telescope was invented.

The difficulty is that, even if there were no expansion, eventually all stars in our own galaxy will run out of nuclear fuel, cool off, and die. Sooner or later everything that could conceivably give us the energy we need to live will run out. That's what's meant when we say that the universe itself will suffer a "heat death." That assumes we have the physics right, which, as you say, is not necessarily a well-founded assumption.

PAUL: Well, if we can't be sure about the ultimate fate of the universe, can we at least say something with any more certainty about the fate of our own planet, or our own solar system?

GUY: Actually, in that case, yes. It's safe to say that the ultimate end of planet Earth is inevitable. That's a simple observation of astronomy, something astronomers like us at the Vatican Observatory can talk about with a certain expertise.

To begin with, we've started tracking more and more small asteroids, and it turns out that a lot of them come close to Earth. I follow a Twitter account from the Minor Planet Center in Cambridge, Massachusetts, and it seems like every week they announce yet another near-miss of some chunk of asteroid coming by the Earth . . . though how "near" is a "near-miss" is rather loosely defined. Certainly every month somewhere on Earth, someone will report a *bolide*: a meteor-size lump of space rock burning up as it enters our atmosphere and creating a fireball as bright as the moon.

In fact, recently—February 14, 2013—both of those things occurred on the same day. On that day, a lump of rock about the size of a bus fell to Earth from outer space and just missed the city of

Chelyabinsk, Russia; the explosion when it hit the atmosphere blew out windows and injured a thousand people. Fourteen seconds' difference in its arrival and it would have hit the town itself. Meanwhile, on the very same day another asteroid, this one the size of a small building, flew so close to Earth that it actually came closer than where our communications satellites orbit. The two objects probably had nothing to do with each other—they were on different orbits and apparently came from different parts of the asteroid belt. Most scientists think it mere coincidence that they both arrived on the same day.

And neither one was "The Big One." That's still to come. Sixty-five million years ago, a cosmic collision wiped out the dinosaurs, and there's almost certainly another comet or asteroid out there with our number on it. It could hit tomorrow. Or it could hit in a hundred million years. We can predict with statistical near-certainty that it will happen, but we can't predict when.

When people ask me what to do to protect ourselves, I give the same advice: quit smoking, and wear your seat belt. Smoking and car accidents are far more likely to kill you than a rogue asteroid.

PAUL: Which isn't to say that the effort it takes to keep track of potentially hazardous asteroids isn't worthwhile. Maybe if the dinosaurs had had a space program, they would still be here today.

An impact like the one that wiped out the dinosaurs would again disrupt the atmosphere and alter the ecosystem, but presumably planet Earth would still survive, and full of life, at that. We mammals survived the shock that killed the dinosaurs, and we flourished by filling the niche in the ecosystem where the dinosaurs once lived. So, presumably, if the next one wipes us out, we'll be replaced by intelligent cockroaches or some such thing.

But what can astronomy say about the end of our planet Earth itself, or the end of the whole solar system? How much do we actually know, and what is mere speculation?

GUY: Ah; to figure that one out, we can start by looking at our sun, the star we orbit. We know pretty much how it works from observing it and other stars like it, and in labs here on Earth we can also observe the nuclear reactions that provide the energy that makes the sun and the other stars shine.

We know that stars are powered by nuclear reactions that release energy when hydrogen is fused into helium, and helium into carbon, and carbon into heavier elements. But there's a limit to this process. It turns out, the bottom of the energy pool comes when you've fused all the lighter elements into iron. Beyond that, making heavier elements out of iron actually starts to consume energy. (That's one reason that elements heavier than iron, like gold and platinum and uranium, are relatively rare.)

Judging from what we know about how fusion reactions take place—the stuff of hydrogen bombs, or more simply the outcome of certain reactions inside nuclear reactors, where we can actually produce this energy ourselves and see how it works—it's pretty straightforward to predict the different stages a star goes through from the time it first forms, to when it settles down to a steady shine, to when it goes through its last gasps as the fuel runs out in its core, where the fusion takes place.

What's more, we can see all the stages predicted by our theories actually taking place in stars close to us. Since we see it happening to other stars, it's reasonable to assume that the time will come when our sun's core will likewise have gone through its supply of nuclear fuel. When the core can no longer produce energy, it will cool off and contract; then the mass of gases above the core will come crashing down, bounce off the core, and puff themselves out into a cool, dull-red cloud. Judging from the size of the "red giant" stars we can see nearby, once our sun enters its red giant phase, it will almost certainly envelop Earth itself. According to recent calculations, Earth will survive—by the time the solar gases puff out into a sphere large enough to include our planet, they'll be pretty tenuous, and Earth may just keep orbiting around inside the red giant star. The planet itself may survive, but the same can't be said for anything living on its surface.

PAUL: When will this occur?

GUY: About five billion years from now.

PAUL: So you have time to get to confession before it happens.

GUY: Once the sun's dead core cools off, it'll be nothing but an inert, iron-rich ember. Well, helium-rich. You actually need a bigger star than ours to fuse helium into the other elements up to iron; our sun never gets past helium. Of course, the outer gases of the sun (and other stars of its generation) will get spewed into space at this time, and so they will be available to make new stars.

PRIMO PIATTO

PAUL: Ah! Here comes the first course. Strozzapreti pasta.

GUY: Ah, yes, bits of noodle all twisted up, in a tomato sauce. Um, doesn't that name mean "strangle the priest" in Italian? You're the only priest here!

PAUL: It's the end of the world, anyway, so what do I have to worry about?

GUY: Just as long as none of it touches the antipasto. You know what happens when pasta encounters antipasto . . .

PAUL: You deserve annihilation, for a terrible pun like that.

GUY: Don't blame me. I swiped it from my friend Brian Malow, who actually makes his living as a science comedian coming up with jokes like that . . . and performing in cheesy restaurants like this one.

Hmm . . . twisted noodles in red sauce do look something like the tangled magnetic-field lines around a dying red giant star. I guess it ties in with our discussion of the red giant phase of our sun. Certainly the way I spew red sauce on my shirt whenever I eat pasta fits into that theme . . . pardon me while I move away from the table for a minute and clean myself up.

PAUL: Maybe by the time our sun turns itself into a red giant and consumes the inner planets, our human descendants on Earth could fly off in a spaceship to escape its fate. Just as you could change your shirt, we could exchange our planet for a new one around a younger star once ours is gone. But even if we manage to run away, the time will come when that younger star will die, too; the process will continue, over and over again, as star after star finally dies. Eventually the whole universe will "run down" and die.

GUY: Right. And that's really what little Alvy ought to be worried about—not the expansion of the universe, but the heat death of the stars. It's not just our life, or the lives of all humans, or the end of planet Earth, or even just the death of our star, the sun. The entire universe appears to be doomed, if you're willing to wait long enough.

But there's another disturbing aspect to the expansion of the universe that Alvy hadn't considered (which is not surprising, since nobody knew it back when Alvy was a child).

Recall our discussion on Day 1 about the Big Bang theory. We have observed that the universe is expanding. The evidence we have from looking at clusters of galaxies very far away in space (and, thus, given the finite speed of the light coming from them to us, showing us light now that left them very far back in time) is that this expansion has been going on continually all the way back to a "time zero" about fourteen billion years ago, when not only everything we can observe today but even parts of the universe we can no longer see were all confined in a very hot, very dense point.

Indeed, the best data so far indicates not only that the universe is expanding, but that this expansion is actually accelerating.

And there's no reason to think that what has been going on since, literally, the beginning of time should change anytime soon, even if we can't prove that this must be the case.

PAUL: And David Hume would agree. His point was not that we shouldn't use science to predict future events, but that we shouldn't think that science tells us with certainty what's going to happen in the future. The best we can do is make predictions based on how the world has behaved in the past—and that's what we should do. But we can't know with certainty that the world will keep acting the same way in the future as it has in the past.

GUY: The reasonable implication of all this evolution, the initial Big Bang and the acceleration of dark energy, is that the universe ought to keep expanding forever, faster and faster.

But then we come up against something called the "light horizon." According to the Big Bang theory, space is expanding between the galaxy clusters; the more space there is between us and some distant point, the more expansion that we're looking across, and so the faster the point we're looking at will appear to be receding from us. If we look far enough away, the amount of expansion will make something appear to be receding from us at the speed of light. Right now, that "far enough away" is estimated to be at 13.7 billion light-years. That's the "light horizon." Any light that comes toward us from beyond that horizon would have to cross a distance in space that is expanding faster than the light can travel, and so that light could never reach us. Thus, we can see nothing from beyond that horizon.

But it is possible that there is more of the universe beyond that horizon. Nothing—no spaceship, no signal, no kind of information—

can travel from it to us, because in order to reach us, that signal would have to travel faster than light. If we can't see it, likewise, it can't see us. Nothing that happens beyond that light horizon can possibly affect anything in our part of space. So far as we are concerned, it might just as well not exist; we couldn't tell, either way.

However, if the expansion rate was slower in the past, as we now think is the case, then there could be stuff that used to be close enough for us to see—that used to be within our light horizon—but isn't anymore.

In the last twenty years we've discovered that the rate of expansion of the universe has been accelerating since the Big Bang occurred. If the rate of the expansion of space continues to accelerate, then stuff that is now moving away from us at a rate slower than light speed, so that we can still see it, will eventually move beyond our light horizon . . . and then we'll no longer be able to see those distant galaxy clusters. The cosmologist Lawrence Krauss has pointed out that if the acceleration continues, and you wait long enough, eventually *every* other cluster of galaxies other than our own will cross the light horizon. We'll no longer be able to see them. We won't even be able to tell whether there is more to the universe than what we can see nearby.

PAUL: And at that point we'd not even have any way to observe that the rest of the universe is expanding. Which makes me wonder if other essential aspects for understanding the universe have already been rendered invisible to our eyes.

GUY: Exactly! So we may not even know what's out there that's going to end all of existence for us.

PAUL: So, let's see if I have this straight. Elements deep inside a star, where the pressures are high enough, undergo fusion and produce energy. Eventually all the elements deep inside are fused as far as they can be (ultimately, if the star is big enough, into iron). When that happens, the energy supply inside the star runs out, and the star collapses. Then it puffs out into a red giant star, spewing gases from the outer, unfused parts of the star into space and leaving behind an iron core. The ejected gases form into new stars, and the whole cycle begins again. This happens time and time again, until space is littered with dead iron-star cores.

As the universe ages, eventually all the gas will get consumed. And as the universe expands, starlight diffuses into the ever-increasing void, spreading its energy thinner and thinner. Eventually, all the energy in the universe will be dissipated through space, and all the mass will be turned into cold lumps of iron. No stars will shine; no sunshine will exist to fuel life.

If you wait long enough, even the nuclei in the cold lumps of dead stars will eventually decay into radiation; even the most stable of the subatomic particles in these inert atoms might finally decay into radiation that becomes more and more feeble as space expands. Eventually, even black holes will dissipate. So one future scenario is that there is no "end" to the universe. We're just left with a thin gruel of subatomic particles decaying into radiation.

But if it turns out that the laws of physics are slightly different than we've understood up to now, or if some of them change with time, then there could be a more well-defined end of the universe. Some folks speculate that the universe might go through an endless cycle of death and rebirth, with a Big Bang followed by a Big Crunch, followed by yet another Big Bang.

SECONDO PIATTO

PAUL: Ah, they're bringing the main course. Here's where we get to the meat and potatoes.

GUY: Roast potatoes in rosemary, in this case. And thin slices of fillet under a "rocket" salad. Mmm! Of course, all this food is going to be the death of me.

PAUL: Don't be so sure. It's remarkable how much better Italian food is for you than the fatty stuff we get back in America. Living in Italy, I eat great but still lose weight. You know the old argument, do you eat to live or live to eat? Here you can do both!

GUY: I'm reminded of the story of the Jesuit who goes to his spiritual guide and asks, "Would it be OK for me to snack while I pray?" The guide replies indignantly, "Of course not! I'm surprised at you! Snacking would distract you from praying!" A week later, the Jesuit returns and asks his guide, "Would it be OK for me to pray while I snack?" The guide smiles and nods, replying, "Of course, my son. What a good and pious desire, to bring prayer into your snacking!"

PAUL: It's amazing what a change of perspective can do for you— what was bad can seem good, and what was good can seem bad!

Speaking of which, let's get back to Woody Allen. He was worried that the universe will expand to the point that it will all just come apart. We've discussed what science has to say on the topic of the possible far-future states of the universe. But what about the other thing Woody says? According to Woody, since the universe is going to end someday, then nothing is of ultimate meaning or value. So instead of worrying about the meaning of life and pursuing "Big Questions," we should just distract ourselves and enjoy life as best we can. The presence of human life is an accident, anyway.

There's a funny inconsistency to Woody's line of thought. He starts from what he takes to be two scientific facts: in the far future, the universe will become so inhospitable that the human race will come to an end; and the presence of human life is an accident. From this starting point, Woody concludes that nothing is of ultimate meaning or value—there's no point to anything.

But wait a minute. If nothing is of ultimate meaning or value, and there's no point to anything, then there's no point to doing science, and there's no point to believing what science has to say. But this means that Woody's conclusion undermines his starting point. He starts out by believing—valuing!—two scientific facts. And from that he concludes that there's no point to believing what science has to say. That's just not consistent. If you want to make use of scientific facts, then you shouldn't go around concluding that science has no real meaning or value.

GUY: Anyone who's done science knows that, even if science doesn't give us ultimate answers, nonetheless science does have real meaning and value.

PAUL: And that's not just because we humans happen to know it. Science is important because it is true—because it reflects (even if incompletely and imperfectly) the way the world really is. And science will remain (imperfectly) true, whether or not humans happen to be present. Even if the universe ultimately proves hostile to human life, and humanity dies out, scientific truths will remain true for as long as the universe exists. And in some respects, science describes not just what is true about this universe, but also what is true about any possible universe. So the truth of science transcends not just humanity but the universe itself.

I'll tell you who it is that proves Woody Allen wrong—I'll tell you who it is that shows that science has real meaning and value: it's Wile E. Coyote.

As I said a few days ago, he's the long-suffering patron saint of scientists. He is doggedly persistent, and he never gives up. After every painful failure, he dusts himself off and tries again. He relies on brain more than brawn—you can see this in his use of all those clever products from the Acme Corporation.

If the world were fair, Wile E. Coyote would have caught the Road Runner by now—sooner or later, one of those gadgets from

Acme should work! But no, every time—every time!—the Acme gadget functions (or malfunctions) in a way that the Coyote does not foresee.

In effect, the Coyote is doing science with all those Acme products. Science proceeds by a systematic process of trial and error. We assume the world behaves in accordance with physical laws that are regular and inexorable. But since we don't have direct access to those laws, all we can do is check into how things in the world behave here and now, under particular conditions—all we can do is trial-and-error testing. And more often than not it's error—in the lab, something unforeseen always seems to go wrong.

GUY: The only way you can succeed in the lab is if you expect failure—or if you are able to see failure as a kind of success. Every time you figure out why something didn't work, it's a step forward—you've figured out how to avoid that problem in the future. But you also expect that there will be more problems and failures. Wile E. Coyote gets this, and he gets it deeply. That's why he doesn't give up. For him, science has meaning and value, even if it goes wrong—no matter what Woody Allen says!

PAUL: It would be so, so satisfying to see the Coyote catch the Road Runner, just once. It all seems so unfair. In one episode, the Road Runner stands atop a piece of rock that is floating in midair. The Coyote holds up a sign saying: *I wouldn't mind—except that he defies the law of gravity!* The Road Runner responds by holding up a sign that says: *Sure—but I never studied law!* The deck is stacked against the Coyote. He has to deal with all the usual problems scientists face in the lab: equipment that won't work, refutations of theory, etc. But beyond that, he has to deal with an adversary who doesn't play by the same rules—one who seems to be immune to refutation by theory.

The Coyote reminds me of a long-suffering research scientist working in a lab. The Road Runner is like an otherworldly figure, the sort you see in Byzantine iconography: existing on a different plane of beauty and glory, neither concerned with this world's troubles nor bound by its rules. The Coyote is trying to get somewhere—the Coyote is trying to get something done. But the Road Runner runs at high speed not to get somewhere, not to accomplish something, but just for the sheer joy of running.

GUY: So, as we've said before, if science is like the Coyote, then faith is like the Road Runner. Science is concerned with how things work in this world. Faith runs all through this world but looks beyond it.

PAUL: And science can't succeed in catching faith and having it for lunch, any more than the Coyote can succeed in catching the Road Runner. Faith plays by a different set of rules. Just as Romeo's love for Juliet was immune to refutation by outside empirical evidence, the love of God is immune to refutation. God's love stands on a rock floating in midair, holding up a sign that says: *I'm the one who sustains law—so I don't need to study it!*

It's not that faith or the love of God somehow authorizes violations of the laws of nature. From the Christian perspective, it is God and God's love that underwrite those laws! But love is real and important. And love is not something that can be captured and encompassed by the methods and categories of science, any more than the Road Runner can be captured by the Coyote.

From the point of view of science, this can all make religious faith and the love of God seem rather annoying and smug. And for that matter, the Road Runner can seem annoying and smug. The poor Coyote wishes that the Road Runner could be captured and encompassed by the laws of nature. But love cannot be captured and encompassed that way.

It is both the glory and the doom of science to try to go too far, to try to explain everything—just ask Galileo! Many proponents of science go too far when they try to explain love, or to explain away God, scientifically. It can't be done. And many people of faith respond with unnecessary fear and panic to these excesses on the part of science—just as some in the Church responded to Galileo with fear and panic.

GUY: So our response to Woody Allen is that science is important, and what we learn from science has real meaning and value—even if our science, like the Coyote, is doomed to constant failure.

PAUL: I return to the theme we started on Day 1 of this conversation: the universe is either a *chaos* or a *logos*; either it is meaningless and pointless, or it is meaningful and rational. We can't prove that it's one or the other; using the notion of "proof" would presuppose that the universe is meaningful and rational! But we can't do science unless we assume that it's a *logos*. Christians choose to live in the hope that the universe is a *logos*—that it is meaningful, valuable, and rational. That is their faith.

GUY: Worth, value, meaning—those things can't be weighed or measured. They are not material. But if you argue that they don't exist, then by the very fact that you find the point worth arguing, you're contradicting yourself.

L'INSALADA

PAUL: Here comes the salad course! I'm having a Caprese salad—tomatoes, *mozzarella di bufala*, basil, and a bit of oil. Simple. Good. Yum!

GUY: It used to confuse me that you get the salad toward the end of the meal in Italy—and also here at the End of the Universe, I see. In America, we're used to having the salad first, to fill you up while the kitchen prepares the main course. But the Italians explained to me that having a salad covered in oil and vinegar interferes with your palate for the wines you'll be drinking with the pasta and main course. So the salad is the capper to the meal, to help with your digestion.

PAUL: Speaking of digestion . . . Should people of faith be able to digest without difficulty what science has to say about the end of the universe? Should people who are into science be able to digest what Christian doctrine and Scriptures have to say on the topic of the end-time? Or is there some sort of irreconcilable difference? In that case, it's time for us to start passing out some spiritual Rolaids.

GUY: The science that Douglas Adams and Woody Allen rely on is up-to-date and modern. But the questions it raises for them about the meaning of life, the universe, and everything else have been around for a long, long time. You can find the same issues raised in the book of Ecclesiastes, dating from more than two thousand years ago, with the famous opening that starts, "Vanity of vanities! All is Vanity!" I like the way that Rabbi Sacks has recast that passage in his recent book, *The Great Partnership:* "'Meaningless! Meaningless!' says the Teacher. 'Utterly meaningless! Everything is meaningless . . . Man's fate is like that of the animals; the same fate awaits them both: As one dies, so dies the other. All have the same breath; man has no advantage over the animal. Everything is meaningless.'"

So Adams's and Allen's concerns about the end of everything don't really arise from science at all. Those questions and concerns have been around for a long, long time—since long before modern science (or even ancient science) came on to the scene.

PAUL: Aren't you going to eat your salad? Can I have it?

GUY: There really is nothing new under the sun.

Here's where there may be a problem of digestion. Our science is progressive. It doesn't stay the same—it's always advancing. When someone poses a question about the world, scientists get to work on it, and the question is refined and refocused. Theories get proposed, tested, refined, and retested. Sometimes theories get abandoned altogether, and new theories are proposed. Sometimes it turns out that the wrong question is being asked, and science moves on to a better question. Science is sometimes messy. But it keeps moving forward, closer to the truth. There's just no comparison between where science was in the days of Ecclesiastes and where it is now.

But it seems that nothing ever changes when it comes to questions about the meaning of life, the universe, and everything else. Philosophy and religion don't seem to be progressive—they don't seem to advance at all. We're still asking the same questions that were being asked 2,500 years ago. Some people think that this means that science is superior to philosophy and religion—that philosophy and religion should be discarded because they are slowing us down in our search for truth.

PAUL: Pope Benedict XVI put forward an interesting take on that question in his encyclical letter *Spe Salvi* (Saved in Hope) in 2007. Benedict points out that the physical sciences are the only area of human knowledge that shows long-term, cumulative progress. He says that long-term, cumulative progress is not possible, and should not be possible, in the area of ethics. And he sees this as a feature of ethics, not as a bug in the system! What Benedict has to say about ethics being nonprogressive can be applied more broadly to philosophy as a whole—and in particular to questions about the meaning of life, the universe, and everything else.

GUY: So why is it that science is progressive, but ethics is not?

PAUL: Here's what Benedict has to say:

> . . . in the field of ethical awareness and moral decision-making, there is no similar possibility of [progressive] ac-

cumulation [of knowledge] for the simple reason that man's freedom is always new and he must always make his decisions anew. These decisions can never simply be made for us in advance by others—if that were the case, we would no longer be free . . . Freedom presupposes that in fundamental decisions, every person and every generation is a new beginning. Naturally, new generations can build on the knowledge and experience of those who went before, and they can draw upon the moral treasury of the whole of humanity. But they can also reject it, because it can never be self-evident in the same way as material inventions. The moral treasury of humanity is not readily at hand like tools that we use.

GUY: Ah, I see. Once a scientific discovery has been made, the genie is out of the bottle—what once seemed puzzling and mysterious becomes obvious and taken for granted, like a tool readily at hand. Scientific discoveries get built into our technology and our way of life, to the point that we don't even notice them anymore. Whenever you use a wrench, you are presupposing the truth of scientific discoveries made by Archimedes. Whenever you flip a light switch, you are taking for granted the truth of science done by Faraday, Tesla, Maxwell, Edison, and others. Whenever you use a GPS device, which depends on knowing a satellite's position and time to a higher precision than Newton's physics could accomplish, you are accepting along the way the truth of Einstein's Theory of General Relativity.

PAUL: Exactly. Of course, you are free to claim that you refuse to believe this or that well-accepted scientific discovery from the past. But sooner or later you'll probably end up contradicting yourself in practice. The results of past science are woven so seamlessly into our technology that it is very difficult to avoid relying on them in the practical order of things. If you really want to refuse to believe that some well-accepted scientific discovery is true, it's not enough to say, "I don't believe it." To be consistent, you'd have to refrain from making use of any technology that has that particular discovery "built in." And that'd be hard to do. To avoid using basic technology, you'd pretty much have to disengage from human society altogether. So, well-accepted scientific truths get "built in" to everyday life, via technology, to the point that we take them for granted without even thinking about it. That's how it works for questions in natural science.

Pope Benedict's point is that it does not (and should not) work that way for "Big Questions" concerning what is good, what is beautiful, and what is the meaning of Life. Answers to those questions should never get "built in" to everyday life to the point that they are so obvious that they are taken for granted.

Why not? Answering the "Big Questions" involves making a choice—it involves human freedom.

Faraday, Edison, and others figured out the science needed to make electric lights work. They've done the work for us, and we don't need to do it for ourselves if we don't want to. The science is "built in" to lightbulbs, and we can use lightbulbs without needing to understand the science that makes them work. But when it comes to knowing what is good or what is beautiful, no other person can do the work for us. Each generation must freely answer, for itself, what is good, what is beautiful, and what is the meaning of life. We can draw on the wisdom of previous generations—but we must decide for ourselves, in freedom.

To sum up: Science is the only area of human endeavor in which cumulative progress occurs. Science accomplishes this by limiting itself to a relatively narrow focus on questions concerning what is true about the natural/physical world. There is no cumulative progress from generation to generation on the "Big Questions," because answering them involves human freedom in a way that accepting the results of science does not.

If human freedom is to be preserved, answers to the "Big Questions" about goodness, beauty, and meaning can *never* be taken for granted—they can never get "built in" to human society the way that scientific answers get "built in" to technology. A society in which the answers to the "Big Questions" are "built in" and taken for granted is a society that seeks to suppress freedom of thought and choice. It is totalitarian.

GUY: The scientific discoveries embedded in technology get taken for granted until someone tries to use that technology in a new application, and it doesn't work. Then it's back to the drawing board!

PAUL: Fair enough. That does happen in some cases. But it hasn't happened yet with wrenches and lightbulbs and lots of other technological devices that we use every day. And it's not going to happen for the vast majority of them, because we've got the science right—or, at least, right enough.

Some people want to say that, because science is progressive, scientific knowledge should have pride of place over other kinds of knowledge that are not progressive. Some even say that only scientific knowledge should count as knowledge at all! To my mind, that's sort of like a mathematician refusing to acknowledge the existence of the irrational numbers. Or it's like someone looking at the world through blue-filtered glasses and declaring that red isn't real.

GUY: So what does all this mean for the question we've been discussing: the end of the universe? Is dealing with the question of the end-time a job for science or a job for religion? Or do both science and religion have claims on this question?

PAUL: Well, science certainly has a dog in this hunt. The question is, what kind of dog is it? Like we said earlier, science is our best bet when it comes to predicting and describing the possible far-future states of planet Earth and of the universe. But science isn't of much help when it comes time to figure out "What It Means" that our world and universe will someday come to an end.

GUY: Science may not be able to tell us "What It Means." But it certainly has tried to figure out what to do about it.

Consider one way in which technology can alter an end-of-the-world scenario. We can track and detect life-threatening asteroids approaching Earth; and if one is detected, perhaps we can deflect that body, or at the very least make preparations to survive the effect of its impact. In that way our science and technology would alter an event that otherwise would occur in the natural state of things.

Granted, I cannot outline a means of how we can prevent the sun from entering a red-giant phase or delay the heat death of the universe—not from any conceivable extrapolation of today's technology. But then, a thousand years ago no one would have been able to understand even the concept of altering the orbit of an asteroid. Who is to say what we or our post-human descendants will be capable of choosing to do in a billion years' time?

In fact, lots of "transhuman" technobabble solutions have been suggested for dodging the end-time. For example, when the end of the sun and Earth draws near, we could get into a spaceship and fly away from our dying sun to a new star. And then, when that star dies, we could move on to another one, and another one, and another, until we run out of stars.

Or here's another possibility: about thirty years ago, the cosmologists John D. Barrow and Frank J. Tipler suggested that human intelligence could endure in the form of intelligent robots after the universe becomes inhospitable to human life. The plot of one of Isaac Asimov's short stories, "The Last Question," turns on a solution of that sort to the problem of the end-time.

Another possible solution could come from theoretical cosmology, which suggests that our universe is just one of a large or even infinite number of universes in the "multiverse." Maybe it will be possible for us to jump over to another universe when ours runs down. Or maybe we'll be able to transmit the "essence" of our being to another universe, in the form of information.

But any kind of "fix" like this just puts off the inevitable until this universe runs down, and then the next, and then the next. And, more fundamentally, it traps you into the murder-mystery model again. Who says that there's any meaning to be had for our lives, if people do remember us at time-equals-infinity? So what? Watch the "Show at the End of the Universe," and order an espresso.

PAUL: Those are all Coyote suggestions: keep on trying new technological fixes, more and more complicated, in the hope that eventually we'll be able to catch the elusive Road Runner—in the hope that we'll be able to survive beyond the end-time. But isn't it possible to approach the end-time in some other way, instead of just as a problem to be solved or fixed?

Some years ago, I was visiting for a few days with a married couple, Tom and Rita. They are dear friends of mine who live in the Bay Area. At that time, Rita was working at an elementary school, teaching kids in the earliest grades. During my visit, Rita came home from work shaking her head about what had happened that day at school. She poured generous drinks, one for her and one for me, and she told me the story. During recess Rita spotted one of her students, Margaret, sitting by herself and crying her eyes out. Rita sat down beside her, gave her a hug, and asked her what was wrong. Out came Margaret's sobbing tale of woe: Her mom had yelled at her that morning. She forgot to bring her glasses with her to school. She lost her lunch bag. Her friends were teasing her. Her jacket wasn't warm enough, and she was shivering with cold. And during recess she had managed to slip and fall, skinning her knee and putting a gouge in her pants.

Rita gave Margaret another big hug, wiped her tears, and said, "It's OK, honey. You're just having a bad day." But Margaret replied, "I know, but I've never had one before. So it's not OK!"

That stopped Rita cold. Usually her hugs are good enough to solve any problem. But not this one. She took Margaret into the school and helped patch up her skinned knee. By the time she got home from work, she realized that she had come up against one of the "Big Questions."

What is it like when you have a bad day for the very first time? What is it like when a maternal hug is just not enough? Neither Rita nor I could remember our first-ever bad day. Adults get used to the idea of having bad days—they're just part of life. When we have a bad day, we expect that better days will follow. That expectation is based on past experience: we've been through bad days before, and things have always gotten better eventually. But if you're in Margaret's shoes, you don't know that. If it's your first-ever bad day, how can you possibly be ready—how can you possibly cope? It might as well be the end of the world.

The end of the world, if and when it comes, will be unlike anything we've experienced before. It will be as though humanity is having a really bad day for the very first time. So all of humanity will be in Margaret's shoes. How can we possibly be ready for that? How can we cope—how should we respond?

Science would see this as a challenge—as a problem to be fixed via some sort of technological solution. But what if it can't be fixed?

Rita and I sat side by side for a long time, sipping our drinks in silence. I tried to imagine some sort of cosmic Super-Rita showing up at the end of time to give humanity a big hug and say, "It's OK. You're just having a bad day." But though I love Rita dearly, that just wouldn't be enough.

GUY: The Jesuit poet Gerard Manley Hopkins wrote a poem to another little girl named Margaret: "Spring and Fall: To a Young Child." It's one of my favorites. I memorized it back when I was in graduate school and drove my fellow grad students nuts by reciting it at them all the time. (Pro tip: you have to read Hopkins's poetry out loud—you'll miss something vital about if you just read it silently to yourself.)

Here's how it goes:

> MÁRGARÉT, áre you gríeving
> Over Goldengrove unleaving?
> Leáves, líke the things of man, you
> With your fresh thoughts care for, can you?
> Áh! ás the heart grows older
> It will come to such sights colder
> By and by, nor spare a sigh
> Though worlds of wanwood leafmeal lie;
> And yet you will weep and know why.
> Now no matter, child, the name:
> Sórrow's springs áre the same.
> Nor mouth had, no nor mind, expressed
> What heart heard of, ghost guessed:
> It ís the blight man was born for,
> It is Margaret you mourn for.

Little Margaret is having a bad day like she's never had before. Her favorite tree is losing its leaves! If you've never seen that happen before, it comes as quite a shock and source of grief. But by the end of the poem, Hopkins knows: "It is Margaret you mourn for."

Trying to figure out why trees lose their leaves, why death occurs, or why the universe ends—for science, those are "mysteries" that go away once they have been "solved." Once we figure out the answers, they become part of our toolbox—they get "built in" and become taken for granted. But the mystery of why we mourn is not a mystery to be solved and set aside. It is a mystery to be pondered in our hearts and deepened. In the hands of a poet like Hopkins, it is a mystery that can become beautiful.

PAUL: If you see death or the end of the world only as problems to be fixed or solved, you miss out on the possibility that they might be beautiful—that they might be something to be savored and deepened.

GUY: But wait a minute, be careful with this talk about beauty and savoring. It's suffering, death, and the end of the world we're talking about here. Let's not take the easy way out. There's a certain kind of pious Christianity that would like to reassure us: "Don't worry, it doesn't matter all that much if you die or if this world comes to an end. It won't be all that much of a loss. This life doesn't really matter; this world doesn't really count. Things will be better in the next world—in heaven."

I'm sorry, but that does not work. It doesn't work to say that we don't have to care about death or about the physical end to the universe, since we'll continue on in some sort of "spiritual" way. No way. To use the wonderful words of John Polkinghorne (the physicist/Anglican priest), "We are not apprentice angels." We're not just stumbling around here for no obvious reason, stuck in these stupid, material bodies, waiting for the moment when we can be freed of them and emerge, like Casper the Friendly Ghost, to fly off to some astral plane.

It's not the scientist in me that insists on calling that attitude nonsense. Science has nothing to say on this topic, one way or another. What concerns me is that it's bad Christianity. It seems to me that the Creed is very clear on this. Life everlasting is not just spiritual. We are not promised Jesus Christ only in some purely spiritual sense. Rather, we are told that He will reign in this Kingdom, without end. Granted, we have no idea where or when or how to locate that Kingdom. But it exists. Indeed, we are told that this "Kingdom" is already at hand. How that works, I do not know— not completely. I have no data from the future.

This physical universe has importance, and it has meaning.

Matter matters. God cared for it so much, He sent his only Son to redeem it.

My point is, it doesn't work to introduce God and religion to dodge questions about suffering and death.

Some people seem to think that's what religion was invented for: to do nothing more than comfort us in this world, while we wait for the next. We need to spike that one, too. It is not the point of Christianity to evade the "Big Questions" of death and the end, but to confront them head-on, with meaning.

The funny thing is, often the ones who think that people believe in God for the sake of the "comfort" it brings are those who don't believe in God. And I'm not the only one who's been irritated at that. Francis Spufford comments on this in his marvelous book, *Unapologetic*. He writes:

> Lots of atheists seem to be certain, recently, that [death] ought not to be a problem for believers, because—curl of lip—we all believe we're going to be whisked away to a magic kingdom in the sky instead. Facing the prospect of annihilation squarely is the exclusive achievement of—preen—the brave unbeliever. But I don't know many actual Christians (as opposed to the conjectural idiots of atheist fantasy) who feel this way or anything like it.

I remember having a conversation about the "comfort" we get from religion with a friend of mine at MIT, a fellow Catholic, years ago. "Comfort?" she scoffed. "Comfort? Hell!"

Hell, indeed. If you believe that who you are and what you do has eternal significance, it is anything but comforting. Admitting the possibility that hell might actually exist is not designed to give one comfort.

Lois McMaster Bujold's space opera, *Cordelia's Honor*, has a wonderful scene where the utterly ruthless emperor of Barrayar, on his deathbed, reflects on the horrors he committed to hold on to power (and prevent even worse things from occurring). He confesses to Cordelia, "I am an atheist, myself. A simple faith, but a great comfort to me, in these last days."

PAUL: In *The Hitchhiker's Guide to the Galaxy*, Douglas Adams lampoons the kind of religion that points only to the next life. There's a running gag about the great prophet Zarquon, whose disciples await faithfully his promised but long-delayed return. Toward the end of a meal at Milliways, the Restaurant at the End of the Universe (the same fictional restaurant where you and I are eating right now), Zarquon finally returns in glory, with celestial trumpets blaring. His disciples are ecstatic. But just ten seconds later, the universe comes to an end, with Zarquon mumbling apologies about having been unavoidably delayed.

This is, of course, a delightfully scathing send-up of the Christian expectation that Christ will return in glory at the end of time. Adams is saying effectively: "Thanks for nothing! Why did you wait till the end of time to return? We could have used your help sooner, here and now in this world!"

That's a fair challenge. Christians need to be in a position to say why the hoped-for future coming of Christ is something more than a promise of pie in the sky by and by. As we are exhorted in the First Letter of Peter: "Always be ready to give an answer to anyone who demands from you an accounting for the hope that you have. But do this with gentleness and respect."

GUY: Throughout the *Hitchhiker's Guide*, Douglas Adams constantly points out how small and meaningless Earth is in the larger scheme of the universe. The destruction of Earth ends up being no more significant than the demolition of a house to make way for a highway bypass.

Ouch.

But if Earth is just an insignificant dot in an unimaginably large universe, do we have any right to think of any of our questions as "Big Questions"? Are we being presumptuous if we imagine that any of our concerns about life, the universe, and everything count for a hill of beans in a universe that is large beyond our comprehension?

This question becomes particularly poignant when we think of little Margaret who mourns her tree—or of little Alvy, the stand-in for Woody Allen in the movie *Annie Hall,* who has decided that the universe has no point or meaning if it's going to expand into nothingness someday. What is touching about little Margaret and little Alvy is that they both have yet to realize that what they are really mourning is their own coming death. As we see Margaret mourning her tree, or Alvy mourning a universe that will expand into nothingness, we recognize the fateful moment when they take the first step toward grappling in an adult way with the "Big Question" of their own mortality.

PAUL: Douglas Adams's satire is so effective because he manages to make our "Big Questions" seem trivial and laughable. That is uncomfortable—I wouldn't want to be the one to tell little Margaret or Alvy that their concerns are trivial and laughable. But the funny thing is, humor—especially dark humor—is the most effective genre in which to talk about questions of death and the end. It gets us there so effectively.

I know from my own family what it's like to be at an Irish wake, telling jokes and drinking in the presence of the dead body. I must admit that the genre of eschatological and apocalyptic literature in the Bible seems humorless to me. I think I prefer Douglas Adams when it comes to thinking about the end-time.

GUY: Maybe the Bible isn't as humorless as you think. Humor is the hardest genre to translate from one language to another, not to mention from one culture to another. It's the superposition, the tension between two contradicting meanings, that makes a joke funny.

I'm convinced that the book of Jonah was invented to be a funny story, a comedy about the Bible's worst prophet. Even the book of Job can be seen as dark humor. That's what the Coen brothers did with their movie *A Serious Man*. So maybe someday they should take a look at John's book of the Apocalypse and see if they could make a movie about it as a kind of satire.

Every joke, every movie, every story is more than just what happens at the end. The end has meaning only because of what has gone before. And what goes before, the whole telling of the story, is the source of the joy and satisfaction we experience in a well-told tale. So a joke is more than just a punch line. Indeed, it's more than just words; it's the setup, the delivery, the timing, the pleasure of anticipation. A good joke well told, like any well-told story, can be enjoyed over and over again.

The point of the punch line is to tell us the story is finished. A never-ending story is one we can't enjoy, just as a painting without boundaries or a frame loses its focus, its reason for being. So maybe "The End" is a necessary part of every human life, framing it, making it into a whole, and giving it meaning. Similarly, maybe "The End" is a necessary part of the universe, framing it, making it into a whole, and giving it meaning.

PAUL: Then it's no accident that Jesus's most distinctive and effective teaching in the Gospels happens in the form of parables—strange little stories that always include an odd, surprising tension and twist. In the parables, Jesus is always trying to describe what the Kingdom of God is like. He always tells the stories in terms of things familiar in this world—sheep, shepherds, vineyards, stewards, muggings, wayward children, etc.—but he's always pointing to what is strange and unfamiliar about the world to come, which is already-but-not-yet here.

The parables aren't jokes—they aren't funny. But the endings always knock you off balance in much the same way as the punch line of a joke does. Maybe you have to resort to parable form to talk about the meaning of the end of the world—which for Christians is somehow all at once an end, a beginning, and a continuation.

GUY: The end of the world is coming . . . tomorrow, or a hundred billion years from now, or right now out the window of this restaurant. I don't know what we can do about it. But the fact that we can contemplate it, and that we love to make predictions about it that we never expect to be able to confirm, or, even more, that we can make jokes about it, is itself the evidence that the world that comes to an end is not all that there is to the world. Perhaps, in the end, there are things that do remain. Saint Paul numbers them as three: faith, hope, and love.

DOLCE

PAUL: Ah, time for dessert and coffee! I won't go so far as saying that it's the "End of the World" if I don't get coffee. But things are better with coffee.

GUY: Forget coffee—give me dessert! Throughout the whole meal I have been looking forward to this moment, anticipating it with hope. I'll take a taste of your *tiramisù*, thank you. If you're good, maybe I'll give you a taste of my *tartufo nero*. With dessert at hand, the "End of the Universe" seems much less worrying.

But there's one thing that's still bothering me. I'm satisfied that religion and science can get along with each other on the question of the "End"—I don't see any conflict between them.

Sure, there are going to be Christian fundamentalists who think that science must be rejected, since it makes claims about the end-time that are at odds with a literal reading of Scripture. And there are going to be scientific fundamentalists who think that the Christian faith must be rejected, since a literal reading of biblical accounts of the "End" is at odds with what science has to say on the topic. And there are going to be still others who think that the Christian faith must be rejected because it makes a promise of pie in the sky, by and by; not only is it vain and illusory, since there's no data to support it, but trusting to a vindication in the next world can tend to undermine the struggle for justice in this world. I have sympathy with all of those groups, and I wish them well. But they are wrong. The Bible was never meant to be read literally as a science textbook, and what Christianity promises in the "End" is not pie in the sky, either now or in the by and by.

I'm also satisfied that the scientific investigation of the end-time offers something very positive to the Christian faith. Christians believe that God created the world and sustains it in being; and they believe that the world, therefore, reflects its Maker—it is an intelligible *logos*; it can be understood via reason. Christians also believe that human beings are made in God's image. When we do science—when we try to make sense of the world via reason—we are imitating God; we are acting in the image of God. And to do that is to give praise and honor and glory to God. So when scientists come up with predictions and descriptions of how the world and universe will end, no matter whether or not they believe in God, as far as I'm concerned, by what they do, they are giving praise and honor and glory to God.

But here's what's still bothering me. Is there anything positive to be offered to science from Christian faith—and in particular from what a Christian believes about the "End"?

PAUL: Hope—that's what Christianity has to offer. Christianity is at bottom a certain, peculiar hope that we have in Christ. Even to people who don't believe in God or don't believe that Jesus Christ was the Son of God, that hope is on offer and is effective.

Christian hope is not a naïve or simple optimism. It is not a contrary-to-fact expectation that things will work out OK in the end. Often things do not work out OK in the end. Christian hope is expressed, elliptically and mysteriously, through Jesus's preaching of the Beatitudes. Jesus says that you are blessed when you are poor, when you hunger, when you weep, and when people hate you. He's not a sadist. It's not that he wants us to suffer, and it's not that he thinks that it's somehow good for us to suffer. But as Son of God, he has joined us in our suffering. He doesn't take our suffering away, but through his death and Resurrection, he changes what it means. At least, that's the hope of Christian faith.

When it comes to science and hope, I think again of Wile E. Coyote: now there's a techie-scientist who operates with indefatigable hope, despite the fact that things regularly do not work out. I think of Galileo, five years after his arrest and trial, blind and under house arrest, publishing the *Discourse on Two New Sciences*, his greatest work, the book that showed Newton and those who followed a whole new way of doing physics. I think of the Magi, using what science they knew to follow a star to parts unknown.

Do scientists need Christian hope to be able to continue to work despite setbacks? Nope. Will they produce higher-quality science if they let their lives be animated by Christian hope? Nope. Christian hope doesn't make for science that is "better" or "more progressive." Nevertheless, it changes everything.

GUY: I need more to go on. That's not enough to persuade some of the techie-scientists I work with.

PAUL: It isn't my goal to persuade them. But if you like, here's a true parable:

Remember my friends Rita and Tom, whom I mentioned earlier? Some years after that time when Rita gave a big hug to little Margaret on the playground, I was visiting with Rita and Tom once again. By now they had four little boys, ranging in age from ten on down. I offered to babysit on a Friday evening, so that Rita and

Tom could get away and have some time to themselves. They took me up on the offer and went out for a night on the town, leaving me home with the four boys.

The boys and I had a great time together. I took them out to see a movie. On the way home we stopped for burgers and fries. Back home, we watched a video while playing games and having a bed-time snack. And I still managed to herd them all off to bed by 9:30 p.m., the bedtime specified by Rita.

By then I was pooped. I collapsed into an easy chair in front of the TV, popping open a beer. I am not a parent, and the evening had given me new respect for what parents go through every single day. Wow, the energy kids have and the demands they make! Parents face that every day, and there is no escape!

So there I was, sitting in front of the TV, sipping my beer, feeling relieved to be "off-duty." But after ten minutes I heard little foot-steps coming down the stairs. And there at the door was Danny, who was four going on five.

Rubbing his eyes, Danny declared, "I'm thirsty. I need a drink of water!"

I got him a drink of water and took him back upstairs. He climbed back into bed and seemed to settle down. I tiptoed down-stairs, back to my TV and beer. But a few minutes later Danny ap-peared again at the bottom of the stairs. This time he was visibly upset.

He announced, "There's something scary making noise in the closet. I can't go to sleep!"

I marched him upstairs, and the two of us took a good look in his closet. We didn't find anything scary. I reassured Danny that there was nothing to be afraid of and that everything would be OK. He settled back into bed, and I went back downstairs to my TV and beer. But not ten minutes later, Danny came downstairs again. This time he was crying—big tears rolling down his face.

With a quavering voice, he sobbed, "I can't sleep! There's a mon-ster under my bed!"

I wiped his tears, gave him a hug, and took him back upstairs. We looked under his bed, and we didn't see any monsters. I got Danny back into bed and just stood there looking at him for a moment. And finally I realized what he needed.

Now, any parent would have known much sooner what was needed. But I'm a priest with no kids, so it took me a while to catch on.

I said, "Danny, I'll tell you what. Let's both go to sleep. I'll lie down on the floor next to your bed, and we'll both go to sleep. Does that sound OK?"

Danny broke into a big smile and said, "Yeah!"

So I lay down on the floor, where Danny could see me. In less than five seconds he was out cold, dead asleep.

I was amazed that a human being could fall asleep that soundly, that fast, especially after being so upset.

I lay there for a few minutes, watching Danny sleep. And then, after a little while, to my great surprise, I began to weep. Not sobbing, but lots of tears. At the time I didn't understand exactly why. But as I went back to that moment in memory and in prayer over the following days and weeks, I began to understand.

At one level, my tears were born of sadness and grief. After that evening with the boys, and especially after seeing Danny being able to fall asleep just because I stayed with him, I was feeling more acutely than ever before what I was missing by not being a parent. Don't get me wrong: being celibate and a priest is my vocation—it's right for me. But even so, in that moment I was grieving that I'll never have a child of my own who will look to me every day with the sort of trust and dependence Danny showed that evening.

My tears were also tears of anger. At that time, the scandal involving sexual abuse of children by Catholic priests was just starting to hit the papers and gain public attention. As I saw Danny sleeping there, so young and vulnerable, it enraged me that any priest would ever hurt or abuse a child.

And finally, my tears were tears of joy and hope. As far as Danny was concerned, there was still a monster under the bed, and there was still a scary thing in the closet. We hadn't really done anything to chase them away. But even though those monsters were still present, he was able to go to sleep peacefully, just because I was there with him.

Even apart from the presence of monsters, going to sleep at night can be a scary thing for a little kid. Adults are used to the idea that they'll wake up dependably the next morning. But for little kids, the world is brand-new, and there's not all that much past experience to go on. For them it's not quite so obvious that morning will come, that the sun will rise, and that they'll wake up the next day to new life.

Little kids resist going to sleep at night because it's like facing the "End." If you're a little kid, going to sleep can raise "Big Questions."

Danny was able to go to sleep peacefully that night just because I was there. For all he knew, the monsters would attack overnight. For all he knew, morning would never come. But he was able to go to sleep. It wasn't because of anything special about me, other than the fact that his parents trusted me. That was enough for him— that gave him enough peace and hope to go to sleep, even though he wasn't really sure that everything was going to be OK.

Danny, blessed are you when you are thirsty and ask for a glass of water; you and all who thirst will be satisfied.

Blessed are you when there is a monster under your bed or when there is something scary in your closet. We all have monsters under the bed and scary things in the closet. You will be consoled and kept safe.

Blessed are you when you cry, Danny. Our tears let God know the hurts that we can't yet put into words.

Blessed are you when you go to sleep, even though you fear that everything will not be OK.

For each of us, the "End" will come—the day will come when we go to sleep and do not wake up. And the "End" may come for the whole human race someday. When that time comes, how we go to sleep will be the measure of us.

Christians can go to sleep easily and peacefully because there is One whom they trust who is willing to be there with them and who is willing to die for them. That's what allows Christians to see the "End" not as a problem to be fixed, or a mystery to be solved, but as a culminating moment to be anticipated, savored, and deepened.

IL CONTO

GUY: The "End" is nigh: they just brought us the bill. I hope you brought your wallet.

And, wait a minute . . . weren't we supposed to be watching for something while we were here?

PAUL: What?

GUY: The End of the Universe! We got so busy talking that we completely forgot to watch the End of the Universe out the window! Did it happen or not? I missed it!

Day 6: Would You Baptize an Extraterrestrial?

"WE'D LIKE TO BE THE FIRST TO WELCOME YOU . . ."

PAUL: So, Guy, has anyone really asked you to your face if you would baptize an extraterrestrial?

GUY: More than once. I think the most memorable time I got that question was when I was visiting England to give an astronomy talk at the Birmingham Science Festival, in 2010.

As it turned out, the day of my talk happened to coincide exactly with the visit of Pope Benedict to Birmingham. So the cream of British journalism was there. I had agreed to give an interview to publicize the Festival, but all they wanted to ask me about was the Pope.

PAUL: Understandable.

GUY: Except they kept asking me questions like "What is your biggest source of conflict about the Pope?" Or "Has the Pope ever tried to suppress your scientific work?" Completely out of left field!

They didn't want to hear me tell them how much Pope Benedict supported the Vatican Observatory and its scientific work. So, finally, frustrated that they weren't getting the story they wanted out of me, one of them asked, "Would you baptize an extraterrestrial?"

PAUL: What did you answer?

GUY: "Only if she asks!"

PAUL: I love it! How did they react?

GUY: They all got a good laugh, which is what I intended. And then, the next day, they all ran my joke as if it were a straight story, as if I had made some sort of official Vatican pronouncement about aliens.

PAUL: I like your reply, even if it was meant only as a joke. Really, the question of baptizing an extraterrestrial, an "ET," should come up only if ET really exists and only if ET asks for it, with a good understanding of what baptism means: initiation into the hope and challenge of the Christian way of life—the way of life modeled by Jesus and lived oh-so-imperfectly by Christians ever since. In a sense, the question of whether you'd baptize ET is no different from the question of whether you'd baptize any of the people walking past us in this terminal: you wouldn't baptize someone who didn't want to be baptized or who didn't understand what it means. But, of course, in ET's case there'd be some additional obstacles: Could ET understand our language well enough to grasp what baptism is and ask for it? Would ET understand the actions of washing and anointing, which are so central to the ritual act of baptism?

Speaking of the people around us in this terminal we're visiting today: I'm tempted to think that maybe some of them are extraterrestrials. I've been seeing some pretty strange outfits. But this is California, after all. I'm from Cincinnati, which is deep in "flyover country." So to me, California often seems a bit exotic and alien.

GUY: If you want to find a mixture of more different races and peoples in one place, you couldn't pick a better spot than the Bradley International Terminal at LAX, in Los Angeles. Every family that walks past us is dressed in a different garb, speaking a different language. People are coming and going to Australia, to China, to the Middle East. And this building is great—it's big enough to land a spaceship inside here.

I'm sure you've had an experience like everyone in this airport is having, traveling to another city or another country: once you arrive, you're faced with all sorts of unexpected little oddities . . . the things people typically have for breakfast, the way light switches work, the side of the road the cars drive on. These small changes to our routine make us recognize and appreciate the things we take for granted back home. They help us define ourselves, by showing us what is special and different about ourselves. In the same way, thinking about extraterrestrials is a way for us to reflect on what it means to be human.

But there's a particular reason that we are here in Los Angeles. Although neither of us is from Los Angeles, you and I both lived here for a while during the final stage of our studies, leading up to our final vows as Jesuits.

I lived in a funky neighborhood east of Hollywood, not too far from the central part of Los Angeles. I remember one time waiting in line in a drugstore to get a prescription filled, and in front of me was an elderly Chinese man, clearly worried about his medication. The young pharmacist waiting on him, who was of Chinese heritage herself, was so wonderfully patient, explaining things to him calmly . . . and in Chinese. The next person in line was a harried young Hispanic mother; again, the pharmacist took the time to explain to her the medicine she needed to give her child . . . this time in Spanish. And, of course, when my turn came, she took care of me in English, which was clearly her mother tongue.

Whenever I think of Los Angeles, I think of that young woman at the pharmacy. There's something special about a place where a scene like that can occur and be considered just an ordinary part of an ordinary day—a place where no one is an alien, and everyone is welcome.

PAUL: You say that in California no one is an alien and everyone is welcome. Sometimes it seems to me that in California everyone is somehow an alien! I guess it's a matter of perspective.

To its credit, California welcomes more immigrants than any other US state. But a lot of them do feel very alienated, as they struggle with the language, with poverty, and with their legal status. When you live in California, the question of whether you are an alien and whether you are welcome is more at issue than it is in other states. The issue is clearly on the table.

GUY: So, you're the priest here—you're the one who does some baptizing now and then. Would *you* baptize an extraterrestrial?

PAUL: Ha! This will be fun!

But first, some disclaimers. When I'm representing and acting on behalf of the Catholic Church, in my role as a priest, it's not up to me to decide who can or cannot be baptized. It's up to the Church to decide that. So my first and simplest answer to your question would be this: I'll be ready and willing to baptize an extraterrestrial if and when the Church decides it's OK to do so.

Of course, as Jesuit astronomers, you and I might have opinions on the question that may be better-informed than some . . . though less-informed than others. We've both studied a lot of science, and we've both studied a fair bit of philosophy and pastoral theology as part of our training as Jesuits. But neither of us is a trained academic theologian. Bottom line: what we say shouldn't be taken as any kind of authoritative teaching on the part of the Catholic Church. Fair enough?

GUY: Fair enough.

And there is something else that I want to make clear, right from the start, before we go any further. I want to state, with whatever authority I have as a scientist and as one of the "Official Astronomers" at the Vatican Observatory: Neither I, nor anyone I know, has any evidence that extraterrestrials exist. I do not know of any credible evidence at all that there has ever been contact of any form between extraterrestrial aliens and Earth. Period.

I cannot imagine a circumstance where such contact could be kept secret for very long. And I say this not only as an active astronomer for forty years, but also as someone who knows lots of people in the SETI (Search for Extraterrestrial Intelligence) community who would love to have such evidence—just think of the funding they'd get! And I can speak as someone who's been an officer in the American Astronomical Society and the International Astronomical Union. If there were something like this going on, we'd all be talking about it. There isn't, and we aren't.

Look, I work with meteorites. There was a time when most scientists were skeptical about the idea of "rocks falling from the sky"—and rightly so, since most of the time rocks do not fall from the sky. However, today we have some actual samples that we can work on in our labs (I work every day in such a lab with a thousand such samples in a cabinet not ten feet away from my desk) that we can measure for things like exposure to cosmic rays, things that convince me they really did originate in space. Because of the evidence of these samples, I accept that they are rocks that fell from the sky. Indeed, we have actually observed an asteroid in space that later hit Earth and dropped meteorites—this happened over Sudan a few years ago.

But even knowing that meteorites do exist, I am always skeptical of any rock that someone brings to my lab as a potential meteorite. Virtually all those samples turn out to be terrestrial rocks; not "meteorites" but "meteor-wrongs." Only once has such a piece actually turned out to be a real meteorite. But it did happen, once. And the reason I was convinced in that case was that I had the piece in hand, and I was able to make the appropriate chemical tests that showed it was identical to a well-known meteorite.

How does this relate to the UFO/alien connection? The point is this: We have meteorites. But we have no artifacts from UFOs in our labs. None. Zilch. We have nothing to test. Thus, my skepticism remains.

I also note that there is a population of very skilled observers who spend a lot of time outdoors at night, who are not connected to (or controlled by) the government or any other "official" institution. I'm talking about the large number of amateur astronomers around the world who own their own telescopes and spend hundreds of hours observing the skies. And you find that these people tend to be the most skeptical of all about UFO claims. Because they know the sky really well and spend a lot of time looking at it, they are very familiar with what is out there . . . and though they see many unusual things in the sky, they are able to identify them and explain them without talking about UFOs, extraterrestrials, or other unidentified causes.

PAUL: I'm wondering if it is only relatively recently that people have become fascinated with possible UFOs and extraterrestrials, or has this fascination been around for a long time?

GUY: Steven J. Dick and Michael J. Crowe, two historians of astronomy, have written or edited a number of books on the history of how astronomers have dealt with the question of extraterrestrial life. Just to name two of them: Crowe's *The Extraterrestrial Life Debate* contains articles written by astronomers and others prior to 1915 on the question of whether there is extraterrestrial life and what it ought to mean. Dick's *Life on Other Worlds* carries the subject forward into the twentieth century.

And if you want to know the history of theology's take on this question, an excellent summary and review can be found in Thomas F. O'Meara's *Vast Universe: Extraterrestrials and Christian Revelation*. As far as I can tell, it's a first-rate review of the literature. So, for at least several decades scientists and theologians have been writing seriously on the topic of possible extraterrestrial life. As for how long the question has been present in the popular imagination, your guess is as good as mine.

PAUL: So, what have those scholars had to say on the subject?

GUY: Well, one thing you notice is this: The religious believers tend to say that the existence of extraterrestrials would support their religious faith, and the nonbelievers tend to say just the opposite— that the existence of extraterrestrials would invalidate religious faith. So both believers and nonbelievers tend to see the existence of ET as supporting their respective positions!

Nineteenth-century believers such as the German theologian Joseph Pohle or the English astronomer John Herschel argued that because God is so overabundant in His creativity, He must have filled the universe with intelligent beings, not just us. On the other hand, Thomas Paine (the guy who wrote *Common Sense* and *The Age of Reason* during the American Revolutionary War) mocked Christianity for insisting that either, of all the worlds in the universe, God chose to be incarnated only in ours just because "one man and one woman had eaten an apple"; or else that "the person who is irreverently called the Son of God . . . would have nothing else to do than to travel from world to world, in an endless succession of death, with scarcely a momentary interval of life."

In other words, there's nothing on this topic that has been suggested in recent times that hasn't already been discussed, ad nauseam, for hundreds of years. There's nothing new about wondering about aliens and how they fit into our religion. After all, we've been telling stories about alien races and nonhuman creatures since storytelling began. Just look at all the monsters in the *Odyssey* and other Greek legends!

Even the Bible talks about nonhuman intelligent beings. Besides the angels, there's that odd passage at the beginning of Genesis, chapter 6, that refers to some creatures called the Nephilim and describes these "sons of God" taking human wives. And the Psalms and the books of the prophets are full of references to the "holy ones," those "in the sky," the "morning stars . . . and heavenly beings" who sing praise to their Creator.

Back when we were talking about the Big Bang and reviewing ancient cosmologies, we described all those "planetary intelligences" and "daemons" that the ancients assumed existed in the spaces between the planets. That was the worldview of the people who wrote the Bible; they were perfectly happy accepting the existence of other intelligent beings besides humans.

PAUL: I don't know if this is a new phenomenon or not, but it seems to me that nowadays lots of people actively *want* to believe in UFOs—they are hungry to believe that ET exists. And it's not just because they are wondering about cosmology or about how ET would fit into our theology. I get the feeling that some people hope that there's a race out there that is advanced enough to cross vast distances and visit us—a race advanced enough to be able to show us how to overcome all our human problems. Some people want a savior, and they look to the aliens to be that savior.

GUY: Yeah . . . and how's that likely to turn out? Think of the classic science-fiction movie *The Day the Earth Stood Still*, where the alien comes to Earth precisely to help humankind. It doesn't have a happy ending. And, by the way, haven't we already had a Savior visit Earth? Look what happened to Him. (In that movie, the alien calls himself "Mr. Carpenter," in case the parallel isn't obvious enough.)

PAUL: There's something else that I like very much about your joking reply to the journalists, "Only if she asks." There's a sad and tragic history in the Catholic Church, and in other Christian churches, of forced baptism. Before we talk about whether or not to baptize ET, we need to acknowledge that history.

Baptism can be offered as a gift, but it should never be imposed by force or done under pressure. That goes for the Jews, Muslims, Hindus, Native Americans, and others who have been confronted with forced baptism in the past. And that goes for any ETs we may encounter in the future.

"PLEASE DO NOT LEAVE YOUR BAGGAGE UNATTENDED . . ."

GUY: Fair enough. But the gift of baptism can still be offered, even across boundaries, with due respect and sensitivity.

The issue of baptizing people different from us came up right from the start in Christianity; in its first decades, the young Christian Church had to grapple with the question of whether or not Gentiles could be baptized. And the issue has come up again and again through the centuries, as horizons broadened and Christianity spread.

It came up in a particularly interesting way in the Middle Ages. In those days, people knew the world was round, but the common belief was that there was an impenetrable strip of land (think: the Sahara Desert) separating the northern half of Earth from what was then called the "antipodes," the equally temperate areas presumed to exist just as far south of the equator as Europe was to the north. So the question came up, in a theoretical way, among some theologians of the time: are people living on the other side of the divide in need of Christ's redemption?

Saint Zachary, who was Pope from AD 741 to 752, answered by refusing to speculate concerning any beings that were not actual descendants of Adam and Eve. What mattered to him was to insist on the unity of human beings. As he saw it, the essential point was that all human beings are related to one another, and we share a common ancestry. Thus, no person is superior to another by virtue of race or heritage.

The issue arose again, in a very practical sense, in the sixteenth century. That's when explorers from Europe finally managed to sail both to the antipodes—the regions south of the Sahara—and to the New World. In Africa, Asia, and the Americas they met people who looked different from them.

There was a great debate before the Spanish King Charles V in 1550. Slave traders, supported by the scholar and "humanist" Juan Ginés de Sepúlveda, argued that the peoples found in the Americas were fair game for enslavement, since their "barbaric" culture demonstrated that they were less than fully human.

PAUL: Wait a minute . . . isn't LAX airport, where we are today, located on Sepulveda Boulevard?

GUY: Yeah, interesting coincidence. Same name, different family. I looked it up in Wikipedia: Sepulveda Boulevard in Los Angeles runs up to property that was part of the Sepulveda family ranch, Rancho Palos Verdes, which was granted to them back in the eighteenth century. But that's two hundred years after the debate I'm talking about.

It was a Dominican priest (who was later made a bishop), Bartolomé de Las Casas, who defended the Native Americans. He argued that they had souls and that they deserved to become Christian. The Church agreed. That is what started a massive missionary effort, in which we Jesuits played a major part. Anyway, you could argue that sometimes forced baptisms could be seen as an effort to protect people from enslavement.

The efforts of the missionaries often came into conflict with the European colonists who wanted slaves. The Jesuits had built up colonies in South America populated by natives and escaped slaves, colonies called the Reductions. In this they were bitterly opposed by the slaveholders. That was one of many factors—there were others, less noble!—that led to the Jesuit order's being suppressed for forty years at the end of the eighteenth century.

PAUL: Jesuits played a significant role in the European age of exploration. Like the travelers around us at this airport, they went to India, Africa, Asia, and to the New World, North and South America, following the original explorers. So the Jesuits were at the forefront in encountering and mediating the Christian faith to other cultures. I don't know if those cultures were as alien to sixteenth- and seventeenth-century Jesuits as ET's culture would be to us today, but they were still plenty alien.

The Jesuits were men of their times. Like us, they shared the biases and limited perspectives of their time. Still, there was something distinctive and extraordinary about the Jesuit mode of encounter with new cultures, something that tended to raise hackles outside of Jesuit circles and back in Rome.

Jesuits tended to be big believers in and practitioners of *inculturation*. They didn't simply come in and try to impose their own culture on the peoples they encountered. Instead, they made serious and successful efforts to learn local languages and to adapt to local cultural practices and modes of dress. They made serious efforts to adapt Christian ritual practices to local cultural norms.

Some historians and analysts see this as nothing more than a pragmatic strategy on the part of the Jesuits—a way to gain trust and acceptance, to further their ultimate aim of imposing European cultural norms and religion. Other historians think that something more creative and astonishing was going on—that the Jesuits were genuinely open to the idea of learning from the cultures they encountered, even to the point of learning something new about the Christian faith.

GUY: That's consistent with an orientation of mind and heart that is at the core of Jesuit spirituality: "finding God in all things." Jesuits, by training and inclination, are open to finding God present anywhere, in any place or situation.

Of course, the Jesuits who encountered new cultures felt they had something positive to offer: they wanted to spread the Word about Christ; they wanted to evangelize. But because of their tendency to "find God in all things," they were open to the possibility that they would find Christ somehow already present in the new cultures, if only implicitly or indirectly. So they were at some level open to the possibility of learning something new about their own Christian faith from the new cultures they were encountering.

PAUL: So the Jesuits' mode of encounter with new cultures involved a genuine if partial reciprocity.

Don't get me wrong; the Jesuits were inclined to want to baptize as many people as possible. They admired Jesuits such as Francis Xavier who baptized people by the boatload. (Xavier's baptizing arm is preserved as a relic to this day in a Jesuit church in Rome.) But the Jesuits were also inclined to allow themselves to be transformed by the cultures they encountered. They were open to the possibility that European culture and its way of appropriating the Christian faith might be changed in some measure by its encounter with new cultures. They were open to the possibility that new insights to the "Good News about God" could be revealed in the encounter with an alien culture.

GUY: That openness is part of what I love about living and working with Jesuits. That particular brand of idealism is part of what attracted me to become a Jesuit to begin with.

But idealism doesn't always play well in the real world. As it happens, lots of times Jesuit efforts to bridge cultures didn't work—or at least didn't work in the long term. Remember the Reductions, the colonies of natives and escaped slaves? A lot of those "idealistic" societies were ultimately destroyed—by gathering the native peoples together into cities, the Jesuits inadvertently rendered them more vulnerable to attack.

No matter how good your intentions, it is very difficult to communicate effectively across cultures. Misunderstandings and unintended consequences are the rule rather than the exception. Just look at all the folks in the airport here trying to buy a coffee from vendors who don't speak their language or take their kind of money. Heck, sometimes it's hard enough to have a sensible conversation at the dinner table with our own families in our own homes. And if it's difficult to communicate effectively across cultures with other humans—with members of our own species—how much more difficult would it be to communicate with ET?

If there are other planets suitable for life; if there is life on other planets; if that life is intelligent and has free will; if that life can be recognized by us, and us by it; if it is close enough that we can have a meaningful conversation, not hindered by a thousand-years' gap as we wait for each message to go back and forth from their star system to ours at the speed of light; if we are able to understand their language correctly, and they ours—well . . . that's a long string of ifs. And if any one of that string fails, then there will be no conversation. If any of that chain of ifs turns out wrong, we'll never know. The issue of baptizing ET will never come up.

PAUL: I wonder what that conversation would really be like. Do you remember the NASA space probes, Pioneer 10 and 11, which were sent to Jupiter, Saturn, and beyond?

GUY: I was writing my master's thesis at MIT about the very moons of Jupiter and Saturn that they were flying past, and I showed up at Arizona to do my doctoral work just as the guys there were analyzing the data the probes sent back. You bet I remember those probes!

PAUL: What I remember about them are the golden plaques they carried. Since those spacecraft were destined to leave our solar system eventually, they carried with them simple messages in picture form that could be interpreted by an alien intelligence. The messages were supposed to give some indication of who and what we are: portrayals of the human form, diagrams of the hyperfine transition of hydrogen, the position of our sun relative to various specific pulsars, and more. The presupposition seems to have been that basic scientific and mathematical information would be our best "calling card" for making contact with an alien race. So suppose that the Pioneer plaques were discovered by some alien culture. What would happen next? Here are a few possible scenarios:

> *Scenario #1:* Perfect communication. The plaques work exactly as planned! The aliens understand everything and send a reply in a mathematical form that we are able to recognize. Champagne corks are popped worldwide as we look forward to getting to know our new neighbors.
>
> *Scenario #2:* No communication. The plaques are a complete failure. The aliens don't even recognize them as an attempt at communication.
>
> *Scenario #3:* Dangerously bad communication. By extraordinarily unhappy coincidence, it turns out that the markings on the plaques correspond, in the language of the aliens, to a particularly vile insult. The aliens get so offended and angry that they attack Earth and destroy it. (For this depressing scenario, I am indebted once again to Douglas Adams's *Hitchhiker's Guide to the Galaxy*.)
>
> *Scenario #4:* Dangerously good communication. The aliens understand the plaques perfectly and think to themselves: "What boring losers these humans are! All they want to talk about is science and math! You'd think they'd tell us something about their art, literature, music, and religion. But if all they think worth mentioning is science and math, there isn't much point in replying to them. We might get trapped into a long, tedious conversation."

Do I think that Scenario #4 is likely? No. But if you think that all we'll have in common with ET is science and math, then there isn't much point in raising the question of baptizing ET. At least, not yet!

GUY: There are lots of great "first encounter" science-fiction stories. (A lot of bad ones, too.) That's one reason I love science fiction; it gives us the scope to ask these questions and test these ideas, posing them in the way that physicists pose thought experiments to try out new concepts.

PAUL: Asking "Would you baptize ET?" would mean one thing if we had, in fact, already made ET's acquaintance. If we were talking about ET not in the abstract but as a concrete, flesh-and-blood (or, um, whatever) individual, then the question would be mainly about ET: what ET believes, ET's way of life, and that sort of thing.

But so far, the only encounters with ET have been in the realm of science fiction. Since we haven't yet made ET's acquaintance, it seems to me that when someone asks, "Would you baptize ET?" the question is really more about us than about ET. It's a question that has the effect of forcing into the open some of our assumptions about what it means to be human and about what it means to be in a relationship with God.

"THOSE WITH US PASSPORTS OR RESIDENCY CARDS, STEP TO THE LEFT . . ."

GUY: Some people treat baptism as if it were like going through customs here at the airport. They see it as a rite of passage—as something that accepts you into a club or community and separates you from those who haven't been accepted.

In one of my favorite Far Side cartoons, a dog is just finishing a zigzag dash across a very busy road, dodging cars that are whizzing by with scant space between them. The dog's eyes are wide with fear, and his tongue is hanging out. By the side of the road, waiting to welcome the running dog, is a group of dogs, one of whom exclaims: "All right! Rusty's in the club!" Once you've gone through the rite of passage, once you're in the club, you are accepted as a peer and equal, and you have new rights and privileges.

PAUL: But to me, baptism and Christian life are not about rights and privileges. They're about self-sacrificing love and community.

Christian life could sometimes involve the equivalent of finding your way across a dangerous, busy street. But you shouldn't have to get across that street alone. Those dogs shouldn't be sitting there waiting to see if Rusty makes it; they should be with him, helping him get across the street!

GUY: So far we've been assuming that ET would *want* to be Christian and would *want* to be baptized—we've been assuming that Christianity is a club ET would want to join. But can we be sure of that? It's not that I doubt the truth and worth of Christian belief. I'm a believer. But I wonder what sort of chance ET would have of being able to recognize the truth and worth of Christian belief.

After all, the forms of Christian practice and worship can be quite different from one culture to the next. Tip O'Neill said that all politics is local. Though there are universal truths and beliefs that are acknowledged by all Christians, we've both seen—just from coming from the Midwest and now living in Italy—how all Christian practice and worship really *is* local, to the point that one culture's practice of Christianity can seem, well, almost *alien* to Christians from other cultures.

PAUL: So, in other words, since there's such diversity in Christian practice, maybe ET wouldn't be able to recognize believers practicing their faith in different ways in different parts of the world as being part of one and the same Christianity. ET could probably recognize all McDonald's restaurants as being part of the same chain, but would ET be able to recognize Christians in different parts of the world as being part of the same "chain"?

GUY: Every time I hear an Irish priest preaching, I wonder how hard it was for the old Romans at the time of Saint Patrick to trust the faith to those barbaric Celts, much less stomach listening to homilies preached by people who had still been painting themselves blue just a generation earlier. And, mind you, I'm half Irish.

PAUL: I've been puzzled by the diversity of Christian practice even within my own family!

My grandmother grew up in a German-speaking community on a farm in rural southeast Indiana. She was one of a slew of sisters. By the time I was a kid, all those sisters were elderly, in their late sixties and seventies.

GUY: Er . . . for some of us, that's not as old as it used to seem . . .

PAUL: At family gatherings, those sisters would always end up sitting together at the dining room table, talking for hours, and their conversation seemed to me like something from another planet. They were just one generation removed from farm life in southern Germany—they were basically Bavarian peasant girls who had grown up in rural Indiana. So they were deeply informed by a peasant Bavarian Catholicism that was foreign to my suburban-American parish experience.

From my perspective, they practiced something akin to Catholic voodoo. For them, the saints were real powers to be supplicated and manipulated. I remember endless arguments as to what should be done with this or that saint's statue in order to bring about a certain desired result. If you wanted to sell your house, you were supposed to bury a statue of Saint Joseph upside down in the yard. If you wanted to ensure good weather, you were supposed to hang some other saint on a certain kind of bush. If you wanted your daughter to get married, you'd hide yet another saint's statue under her mattress. And so on! There were debates about the right way to call upon the power of various saints. And that was part of the fun. But they had no doubt that the power of saints could and should be deployed.

Well, this just seemed absurd to me. I was amazed, horrified, and fascinated. Budding scientist and rationalist that I was, it boggled my mind that I was just two generations removed from a gaggle of women who were doing Catholic voodoo. I sometimes challenged my grandma about this, but the discussion would go nowhere. Grandma knew that the saints did good work, and if I didn't want to get in on the action—well, that was my loss.

This is the same grandma, by the way, who called my dad on the occasion of the first moon landing to ask how it was that the astronauts were able to land on the *side* of the moon without falling off. Grandma was operating under a different cosmology and physics than the one I knew.

GUY: Well, there's that diversity of practices you were talking about! Christianity takes diverse and sometimes strange forms in various cultures, including the form of "Catholic voodoo" that was practiced by your grandma and her sisters.

PAUL: On one hand, I revel in that diversity. Looking around at all the different people in this airport, where they are going and where they are coming from, how they're dressed and how they carry themselves as they move through the crowds, there's no way you could pick out who's Catholic and who isn't. It could be anyone. I love the fact that Christianity is able to take root—to be "incarnate"—in any culture. That is "finding God in all things"!

But there's a western, scientific side of me that feels compelled to purify Christianity of its "voodoo" and magic. There's a part of me that doubts that it'd be right to invite ET into a faith tradition that is so culturally diverse. How could ET possibly make sense enough of Earth culture to be able to separate the hokey magic from the core of the faith?

I wonder if my desire to separate hokey magic from the core of the faith is, itself, a cultural artifact. Part of me would be horrified if ET saw baptism as being some sort of magic. But part of me knows that I grew up formed by a grandma who saw baptism as being some sort of magic. Oh, my divided heart . . .

GUY: You know, it strikes me that the answers we've been giving so far when we were asked "Would you baptize an alien?"—including my "Only if she asks" answer—haven't really satisfied our questioners. That's why we keep hearing the same question, time and again. Maybe we're answering the question wrong because we've been hearing the question wrong.

PAUL: For one thing, that question can be heard in a couple of different ways, with different emphases. It can be heard either as, "Would you baptize *ET*?"—that is, as a question about if ET is someone to be baptized. Or you could ask it as, "Would *you* baptize ET?"—that is, who are *you* to decide to let ET into our club? A similar ambiguity came into play in early Christianity, when Paul disputed with James, Peter, and others as to whether Gentiles, non-Jews, could be baptized.

Should it be a question mainly about us Christians—about whether we're able and willing to include in the Christian community others who seem very different from us? Or should it be a question mainly about the others—the Gentiles and ET—about whether they are able and willing to take on the Christian way of life?

Maybe "Would you baptize ET?" is not the right question. Maybe it's not the right starting point for a discussion. When you start with that question and try to answer it, you're put in the position of being some sort of gatekeeper for God. I don't see myself in that role. I think that God is perfectly capable of doing the gatekeeping without my help.

And in any event, baptism is not an end itself. It is not a prize to be sought, or an honorific. It is a means to the end of inclusion in a certain kind of community, the Kingdom of God.

GUY: So, what are the right questions to ask? What would be a better starting point?

PAUL: If I ever meet ET, long before the topic of baptism ever comes up, I would want to ask some other questions . . . not only of ET but of myself.

Am I willing to share a meal with ET? Jesus is reported to have been regularly at table in lively discourse with a wide variety of people. Those who were at table with Jesus were often people who wouldn't normally sit down together: Jew and Greek, slave and free, tax collector and zealot, male and female. There was something about being with Jesus that overcame differences or antipathies that would otherwise keep enemies from sitting together at table.

And then I'd ask ET whether she's willing to share a meal with me. If ET and I are willing to share table fellowship (or whatever the equivalent of table fellowship would be in ET's culture), then maybe ET and I are already living as if we are in the Kingdom of God.

If I saw ET sick or wounded by the side of the road, would I stop to tend to her needs? In Jesus's parable of the Good Samaritan, that is the standard of behavior and mutual care that is set for inclusion in the Kingdom of God. And then I'd ask ET whether she'd do the same for me.

Am I willing to suffer or die for ET? Self-sacrificing love is another marker of inclusion in the Kingdom of God, and such love was exemplified preeminently by Jesus himself. And then I'd ask ET whether she'd be willing to do the same for me.

If ET and I answer "Yes" to all those questions, then the two of us are already living as if we are in the Kingdom of God, at least regarding how we treat each other. And if ET can answer those questions "Yes" not just with regard to me, but with regard to a wide variety of others, then I'd have to say that ET is living as if she is already in the Kingdom of God. What would baptism mean for ET at that point? Would it serve to ratify and deepen a reality that is already present?

GUY: Just being able to ask ET those questions would presuppose a degree of intimacy that even most human relationships don't have. There are some people I could perhaps die for; there are far fewer I would be comfortable asking to die for me!

PAUL: Well, Christianity isn't always about being comfortable.

Let's take things a step further—let's try to look at things from ET's perspective. Suppose it had been ET asking me those uncomfortable questions, and suppose I answered yes. That could lead ET to conclude that I, though an alien (from her perspective), am living as if I am already in the Kingdom of God (as understood by ET). In that case, ET might ask me if I would like to receive ET's equivalent of baptism. Then the question for me would shift from "Would I baptize an alien?" to "Am I willing to be baptized by an alien?"

What should I do at that point?

As a Jesuit who is inclined to "find God in all things," I am thrilled when I find God somehow already present in unexpected places—much as the early Jesuit missionaries were thrilled to find God somehow already present in the cultures they encountered in China and elsewhere. But how should I respond when the shoe is on the other foot—when it is ET who is surprised and thrilled to find God somehow already present in me? Should I feel honored, or horrified, or what?

To put it more simply: If we encountered an alien culture, and we recognized that something recognizably Christ-like was somehow already present there, should we be willing to enter into that culture's rite of initiation? Should we allow ourselves to be baptized by an alien?

"TERMINAL MAP: YOU ARE HERE . . ."

GUY: So, is that what you'd have said, if a reporter had asked you, "Would you baptize an extraterrestrial?" Your reply would have been to ask back, "Would you allow yourself to be baptized by an extraterrestrial?"

PAUL: No. I mean, sure, I think it's a good question to consider and contemplate. But that wouldn't have been a good way to answer the reporter.

I like the answer you gave: "Only if she asks!" I'm amazed you had the presence of mind to come up with that answer, right then and there.

GUY: Well, remember the other questions they had been throwing at me: "Has the Pope tried to suppress your scientific work?" and so on. They were pretty aggressive. The reporters were looking for a juicy story and for ways to make me look stupid, or at least to make my Church look stupid. I don't know if they thought I was stupid, but they sure thought that my religion was.

PAUL: Are you saying that you think "Would you baptize an extra-terrestrial?" was some sort of trick question?

GUY: Of course. It was an attempt to trap me in a "gotcha" moment.

If I had just said, "Yes, I would baptize ET," then I would have looked cosmically naïve. I would have been saying that dumb-little-human-me thinks he has the right to preach to highly advanced aliens about what they should believe—aliens so far advanced above any human that they can cross the incredible distances of space to visit us. I would have been saying that aliens ought to give a damn about some human guy who died two thousand years ago, in a dinky little country on our dinky little planet, traveling around our dinky little star, in our dinky little galaxy. If that guy really was the Son of God, then what was He doing wasting His time on our crummy planet? And, for that matter, if He really was the Son of God, why does He need me to do His heavy lifting for Him?

PAUL: That's pretty much the same argument that Thomas Paine used two hundred years ago to mock Christianity.

GUY: On the other hand, if I had said, "No, I would not baptize ET," then I would be admitting that Christianity has no universal or cosmic significance, after all. I would be saying that Christianity is nothing more than a local superstition, amusing for the yokels but not really important in the grand scale of things.

So they thought they had me trapped. But when I simply blurted out the first thing I could think of—"Only if she asks!"—I turned the tables on them.

I made baptism not my decision, but ET's.

If ET, with all her superior technology, decided freely to ask for baptism—if ET, with all her advanced knowledge, accepted that our human Savior really does have importance and meaning for her—then suddenly it'd make the reporters with their petty skepticism look pretty foolish.

And notice how the reporters responded when, quite by accident, I turned the tables on them. They laughed. That was their only possible response—it was the only escape route available to them.

PAUL: In essence, you're answering their question—would you baptize ET—with a different question: did ET ask to be baptized? That changes the ground on which the argument is being contested.

You do realize, of course, that you're not the only guy ever to respond to a "trap" question with a question of your own. That strategy was used to great effect two thousand years ago . . .

When the Pharisees and Herodians asked Jesus if it was OK for Jews to pay Roman taxes, Jesus recognized the trap: if He answered yes, they would accuse Him of honoring a false idol; and if

He answered no, they would accuse Him of insurrection against Rome. So He replied by asking, "Whose image appears on Roman coins?"

The only possible response was "Caesar's."

Jesus then drew the suddenly obvious conclusion, "Give to Caesar what is Caesar's and to God what is God's." The trap was revealed and dismantled, and Jesus's questioners had no choice but to laugh at themselves, lest they look even more foolish.

On another occasion, Pharisees and teachers of the law tried to trap Jesus by asking whether a woman who'd been caught in the act of adultery should be stoned to death. If Jesus answered no, they would accuse Him of disagreeing with the Law of Moses; and if He answered yes, they would say that His teachings about mercy didn't seem to count for much, after all.

Jesus let them stew for a while, and then He asked, "Who among you is without sin? That's who should throw the first stone."

I'm sure there were some stifled guffaws around the edges of the crowd, but those at the center could only slink away silently. But Jesus wasn't done. He turned to the woman who was guilty of adultery and asked, "Hasn't anyone condemned you? Then neither do I. Go, and sin no more." With these questions, Jesus dismantled two traps: the one set for Him by the Pharisees and teachers of the law, and also the trap of sin and guilt that the woman had set for herself.

GUY: I always wondered about that episode. I could imagine that someone overhearing Jesus's interaction with the guilty woman might want to say, "Hey, that's not fair! That's not consistent! Why are you giving her less punishment than she deserves?" We keep asking Jesus to be fair and consistent—especially when it comes to someone else's sins! That's the trap we're always trying to set for Him. If only He'd be "fair and consistent," then we'd know how to manage our relationship with Him—we'd know how to game the system.

PAUL: But Jesus always dodges our trap by responding, in myriad ways, with a question of His own: *Do you love me, as I love you?*

Love is not fair and consistent; it is somehow *more than* fair, *more than* consistent. Don't ask me how or why it works this way, but someone who is deeply in love treats *everyone* well—and one person extraordinarily well. But if all you ask of the one you love is mere fairness and consistency, you're not really in love—and you're not looking to be loved.

So, Guy, those journalists expected that their question "Would you baptize an extraterrestrial?" would reveal you and your faith to be somehow unfair or inconsistent. Your answer turned the tables on them by drawing their attention away from what's fair and consistent and putting the focus squarely on love. When you are willing to wait to hear what ET wants, you're showing you love her. When you love ET, you are *more than* fair and consistent.

GUY: OK, I weaseled my way out of a tight one there. But I think it's time now to address head-on some of the assumptions those reporters were making.

I believe that Christianity is more than just a local superstition for us yokels. I'm betting my life on that belief. (And, anyway, who wants to think of himself as a yokel?) But still, there's the worry: is it presumptuous of us humans to believe that Christianity is something of cosmic significance—that it's something more than just a local fad here on planet Earth? That's really what those reporters were asking me.

PAUL: Or to put it another way: How could it possibly be fair and consistent on God's part to come down to planet Earth and live and die in human form? Shouldn't the God of the entire cosmos bestow attention equally throughout the whole universe, rather than focus in on this little cosmic backwater where we happen to live?

Theologians say these questions are part of what they call "the scandal of particularity."

GUY: So, do you have an answer? Do your theologian buddies?

PAUL: Nope. Not me, not the theologians. And, in fact, God actually does not give an answer to these questions.

Instead, God poses a question in return: *Do you love me, as I love you? Why are you worried about whether I'm fair and consistent? If you'll love me as I love you, you'll see that I'm more than fair and more than consistent. I am Love.*

GUY: Instead of thinking we're so tiny that God couldn't possibly find us and love us, the fact that He actually does find and love each one of us, individually, and gives each of us all His attention, shows just how big God is. God doesn't care about humanity; He cares about individual humans. Like us standing in front of that Seurat painting, He's able to flip back and forth between the individual dots and the whole picture.

PAUL: Christians believe that the God of all things, the God of the entire universe, is in love with us humans. And when you're in love, you show a kind of special interest. That doesn't mean God can't also be in love with other intelligent beings on other planets. Being loved makes you special and unique, even if there's nothing inherently special and unique about you, and even if others are loved, too. That's the "more than" of love; that's its special power.

GUY: I'm a believer; your talk about the "power of love" helps. But still . . . I know my techie friends. They get their hackles up with the idea of God forming a special covenant with a certain subset of Chosen People, or sending Jesus Christ as Savior at a particular time and place in human history on planet Earth. I can see what you're getting at when you focus on love rather than on fairness or consistency. But I'll tell you what I hear all the time:

"Our posturings, our imagined self-importance, the delusion that we have some privileged position in the universe, are challenged by this point of pale light. Our planet is a lonely speck in the great, enveloping cosmic dark. In our obscurity—in all this vastness—there is no hint that help will come from elsewhere to save us from ourselves." That was Carl Sagan, reflecting on a picture of Earth taken by the Voyager 1 space probe from a distance of 3.7 billion miles, in which our planet appears as a tiny pale blue dot.

Or again: "Space is big. You just won't believe how vastly, hugely, mind-bogglingly big it is. I mean, you may think it's a long way down the road to the chemist's, but that's just peanuts to space. No, really!" That, of course, is Douglas Adams, near the beginning of *The Hitchhiker's Guide to the Galaxy*. A later section of the *Hitchhiker's Guide* describes a device called the Total Perspective Vortex: a torture device that drives you insane by forcing you to perceive the full immensity of all of space and time while at the same time showing you a flashing dot with the label *You Are Here*—drawn to scale.

The working hypothesis of modern cosmology, its Cosmological Principle, is that there should be no location in space or time that is special or privileged in any way. The large-scale physical properties of the universe, it is assumed, should look pretty much the same from the perspective of any observer, at any place or time. And people want to apply that principle to everything—life, the universe, everything.

These techies are guys, after all, who often tend to have pretty rough experiences with their own personal lives. A lot of them have a hard time believing that anyone could possibly be in love with them; it violates their own personal cosmological principle, and maybe it's not what they might have experienced themselves. Indeed, when love does happen, they're suspicious. So talking about God as Love doesn't work for them. For them, love is vaporware. Science works.

PAUL: Science is very successful. And since science deals only in terms of energy and matter, it can seem obvious to people who are into science that energy and matter are all that there is—there is nothing else.

As Pope Benedict points out in the encyclical letter we quoted before, *Spe Salvi*, science is so progressive precisely because of how it limits itself to dealing only with the objective "outside" of things—with energy and matter. When you are using the methods of science, everything that is important about your subjective experience of yourself—your freedom and personhood, your personal experience of your body, thoughts, feelings, and so forth—is an irrelevant distraction to the problem at hand. As far as the methods of science are concerned, these things simply don't count and don't need to be taken into account.

But the "problem at hand" that science is good at dealing with isn't the only problem that matters.

It's a mistake to assume (as many people seem to do) that if something doesn't count as far as science is concerned, then it shouldn't count, period. If you think along those lines, you end up in the very odd position of thinking that your own personhood—your most intimate and personal experience of yourself—doesn't count and doesn't need to be taken into account.

People who think this way are trying to take science very seriously and with integrity. They are trying to be consistent. To them it perhaps seems to be a sad but unavoidable consequence of science, true even if difficult to accept, that things that we take to be at the very center of our being—our freedom, personhood, and personal thoughts and feelings—just don't matter or count when you look at things objectively, consistently, and scientifically.

In a very dear and odd way, people who think that way are in a position parallel to that of Christian believers! If you are a scientific materialist of that sort, you tell yourself that science compels you to believe that your personhood is ultimately unreal and shouldn't really matter—even though your common sense insists that that can't possibly be true. And if you are a Christian, you believe in faith that your personhood is so real and important that it matters even to the God of the entire cosmos—even though your common sense insists that that can't possibly be true.

For Christian believers, common sense has trouble letting go of the taken-for-granted conviction that ultimately God must be fair and consistent. It's not that God is unfair or inconsistent. Rather, God is *more than* fair and consistent, because God is Love. It's difficult to put it into words what that means, but anyone who has been in love, or who has been loved, has a pretty good idea.

For scientific materialists, common sense has trouble letting go of the taken-for-granted conviction that, ultimately, reality is just energy and matter. But no, reality is more than energy and matter, because reality is . . .

Well, what is reality?

GUY: Yeah. It's not just the scientific materialists who have trouble with that one. The materialistic worldview is so pervasive in modern culture that nearly everyone has trouble thinking clearly about what, other than energy and matter, could count as being "real."

PAUL: Are we stuck, then? Or is there some other alternative that common sense might be able to accept—some other way of thinking about what is real?

Well, let me take a crack at it. I'm going to start with a couple of basic ideas from Pierre Teilhard de Chardin, a twentieth-century Jesuit paleontologist and theologian. Using those ideas as points of departure, I'm going to try to address our problem here.

One of Teilhard's great books was entitled *The Human Phenomenon.* My first point of departure is Teilhard's insistence that the human being should be treated as being completely open to scientific analysis. Teilhard didn't want to accept a way of thinking that was dualistic—one in which science deals with the objective "outsides" of things—things describable in terms of matter and energy alone—but not with the subjective "insides" of things—such as our internal experience of thinking, feeling, etc. Teilhard proposed that our notion of science should be expanded, to let it include the study of *all* phenomena, even those we think of as being merely subjective. He referred to humanity as the "human phenomenon" to stress the point that everything about humanity should be open to investigation by science.

GUY: I see what you mean. By excluding everything but what you call energy and matter, you're removing all the essentials of what makes a person a person, stowing them behind a barrier as complete—and as artificial—as the barrier that separated our Restaurant at the End of the Universe yesterday from the end of space and time. On the one hand, we have to try to do that in order to be outside, unbiased observers; but on the other hand, we know ultimately that such an isolation is fiction.

And it's more than just the truism from quantum physics that you can't measure anything without altering the thing you measure. It's that there can't be a you who does the measuring, or who even wants to do the measuring, unless you accept that the you doing the measuring is also a part of the you that is being measured.

PAUL: My second point of departure is Teilhard's conviction, from the point of view of his expanded notion of science, that the human phenomenon is the most important phenomenon in the entire universe. Teilhard thought that science itself, properly understood, shows that the entire universe has been oriented toward producing or achieving the human phenomenon, by means of the process of evolution. So as Teilhard saw it, science agreed with Christianity in seeing humanity as being of central importance in the universe. For Teilhard, God's love for humanity was expressed in the fact that humanity is the goal of the evolutionary process of the universe.

GUY: That kind of thinking that we're special, of course, is the part that sticks in the craw of the typical techie.

PAUL: Of course. While lots of regular folks loved reading Teilhard's work, he got clobbered by scientists because of his inclusion of a teleological element in evolution . . .

GUY: Help! Remind me again what "teleological" means?

PAUL: "Teleology" means that nature has some sort of built-in goal. Teilhard saw the process of evolution as having the built-in goal of producing humanity. But for most people who do evolutionary biology, seeing evolution as having a built-in goal is a serious no-no. Teilhard also got clobbered by scientists for giving humanity a special, privileged place relative to the rest of the universe.

But Teilhard was also criticized by Catholic authorities and theologians. He was seen as being too positive about evolution. Back then, many in the Church were concerned that the theory of evolution was incompatible with the Catholic doctrine of original sin.

GUY: Although Teilhard took heat from all sides while he was alive, he has since been quoted with approval even by Popes, including John Paul II and Benedict XVI.

PAUL: To be clear, I don't actually agree with the two ideas from Teilhard that I described above. I think his ideas are interesting, creative, provocative—and in some ways *almost* right. But for my purposes, his ideas are a point of departure. I want to spin his ideas in a way that may give us a path out of our problem.

So here goes. Usually the Christian doctrine that humanity is at the center of God's love and concern is interpreted to mean that God loves humanity in a way that is *different* from how God loves the rest of the universe: if humanity is at the center of God's love and concern, then the rest of the universe is not.

But wait a minute; that's not how love works. Love is not a zero-sum game. Love doesn't exclude—it includes. And if that's true of human love, it's all the more true of God's love, since God *is*

Love. Someone who falls in love treats *everyone* better, loves *everyone* more. When God falls in love with us, God treats the whole universe better—God treats everything in the universe as if He has fallen in love with it. It is in this sense that I would say that God's love is *more than* fair and *more than* consistent.

If we are at the center of God's love and concern, there is *something about us* that God loves. So here's where I want to turn Teilhard's idea inside out: What if, whatever it is that God loves about us is not something that *distinguishes* us from the rest of the universe, but rather is something that we have *in common* with the rest of the universe? What if what God loves in us is something that is utterly characteristic of and typical of the rest of the universe?

GUY: Where are you going with this? What does it have to do with scientific materialism?

PAUL: Earlier we talked about the Cosmological Principle in physics. It's the working hypothesis that there's nothing special or unusual about our particular location in space and time; it'd be presumptuous, after all, to think that we have some sort of privileged location in the universe. Well, I propose that we broaden that way of thinking. Let's adopt the working hypothesis that there's nothing special or remarkable about the human phenomenon—nothing special or remarkable about *us*.

If you buy into scientific materialism, you take it for granted that energy and matter are the only things that exist. But then you find yourself in the odd position of having to think that the "human phenomenon" is a rather odd and special exception in the universe. On that assumption, human thinking, feeling, willing, loving, and so forth, have to be seen as strange and atypical phenomena in the universe. On that assumption, we end up having to see ourselves as being a very strange and special "outlier" in a universe that is nothing but matter and energy.

But isn't that presumptuous? Wouldn't it be humbler, less complicated, and more scientific to start from the assumption that there's nothing particularly strange, remarkable, or unusual about what we humans are, relative to other beings in the universe? Wouldn't that be more consistent with the Cosmological Principle?

We humans are (among other things) material, feeling, thinking, willing, self-aware, free, and loving. If we start from the assumption that there's nothing special or atypical about us, then it follows that feeling, thinking, willing, loving, and being self-aware, free, and material must be characteristic features of the universe, not outliers or exceptions. From this it would follow that what God loves in us, God also loves in the rest of the universe. The reason why we are at the center of God's love and attention is not because of how we're *different from* the rest of the universe, but because of how we're so utterly typical and characteristic of the universe. It is in *that* sense that we're at the center. And I think that's a sense of being at the center that maybe both science and faith could accept.

GUY: George Coyne, the emeritus director of the Vatican Observatory, used to give a lot of public talks with a common theme that "in us, the universe has become self-aware." But, of course, that becomes true wherever self-awareness arises. Maybe the point of the universe is precisely to become self-aware, wherever it happens?

PAUL: Is self-awareness the most important trait that we humans have in common with the rest of the universe? Maybe. Or maybe the materialists are right, and it's matter and energy. Or maybe it's feeling or thinking or willing or loving. I don't know. But if God is Love, and we are made in God's image, then perhaps love is, in some sense, the basic stuff of the universe—perhaps it is somehow the most important thing that we have in common with the rest of the universe.

My point is not to undermine science by making it lose its focus on seeing things as being composed of matter and energy. That's what science does, and science works. But science does not and cannot show that everything is made of nothing but matter and energy. Rather, seeing things in terms of matter and energy is a presupposition that science makes up front—it's how science gets off the ground. I don't want to assert uncritically that love really is the basic stuff of the universe. But I wish that scientific materialists would refrain from assuming uncritically that energy and matter really are the basic stuff of the universe. The question should be left open.

GUY: Carl Sagan was famous for pointing out that "we are star-stuff." But now you're saying that, in addition, stars share in "we-stuff."

PAUL: Maybe the stuff we share with the stars is not just (or even primarily) matter and energy, but perhaps something else—like love.

"THE SHARE-A-RIDE VAN CAN BE FOUND . . ."

PAUL: So now, like travelers standing out here in front of the terminal, baggage gathered around us, looking somewhat dazed, let's look back at and examine one final time this strange "trip" we've taken.

We've approached the question of baptizing ET from so many different angles, it makes my head spin. We've discussed various possible criteria or rules that could be applied to decide who can be baptized—and whether ET can be included. And we've asked ourselves whose job it is to apply and enforce those rules and criteria.

As we've discussed these questions, lurking in the background has been the uncomfortable realization that Earth is a tiny, tiny backwater, relative to the immensity of the universe. On the cosmic scale, from the Big Bang to the heat death of the universe, there's no objective sense in which Earth is at the center of things, or is even important or significant. From ET's perspective, it might seem presumptuous and laughable for us humans to be worried about whether or not he-she-or-it can be welcomed to baptism.

Sure, we can believe that Earth really is a special and unique place, chosen by the God of the universe and privileged over all other places and planets. But then we may wonder why intelligent races on other planets have been left out of the "Good News" . . . or find ourselves faced with the daunting task of spreading the "Good News" throughout the entire huge universe—and hoping to be able to do so without falling prey to the same sorts of errors and biases that sometimes led European missionaries to treat their converts in ways that were paternalistic, unjust, or cruel.

On the other hand, we can just give up believing that there's anything of cosmic importance about Earth, about the human race, or about God's revelation to the human race. We could join with friendly and enlightened ETs in smiling benignly on Christianity and baptism, seeing them as amusing native customs that happen to be meaningful and important to some of the locals here on planet Earth, and give up on the quaint belief that the God of the whole universe decided that Earth was something special and took up residence here in human form.

GUY: Of course, my experience in science has always been that when we find ourselves backed into a corner in this way, facing a choice where neither alternative can possibly be correct, the chances are that we're asking the wrong question.

PAUL: Or, at least, it means that we're coming at the question from the wrong angle. When I was talking before about some ideas from Teilhard de Chardin, I was trying to show how scientific materialists may be asking the wrong question, or operating under the wrong assumption.

But more broadly, when we find ourselves backed into a corner, it's usually because we're coming at things from our own point of view, rather than from God's.

From a human perspective, it seems that the important questions about baptism, salvation, and the Kingdom of God are Who is in? Who is out? How can we tell? Who gets to decide? But I think that from God's perspective, the questions are different. In the Gospel of Matthew, Jesus talks about who will be in and who will be out of the Kingdom of God—who will be the sheep, and who will be the goats.

Notice the criteria that Jesus uses for the selection: He invites the sheep to enter the Kingdom of God, saying: "I was hungry, and you gave me something to eat, I was thirsty, and you gave me something to drink, I was a stranger, and you invited me in, I needed clothes, and you clothed me, I was sick, and you looked after me, I was in prison, and you came to visit me." The sheep reply: "When did we do these things for you?" And Jesus replies: "Whatever you did for one of the least of these brothers and sisters of mine, you did for me."

Does it matter, for baptism, whether you're a human being, whether you happen to come from planet Earth, whether you're intelligent and rational, or whether you happen to be a member of a Christian community? Yes, these things do matter. But it seems that, from Jesus's perspective, they matter less than other things, such as feeding the hungry and visiting prisoners.

Should ET be baptized? Well, that depends on whether ET is, or hopes to be, a citizen of the Kingdom of God. And that depends on how ET treats, or hopes to treat, the least of her brothers and sisters.

Recall again the scenes of Jesus dining with Jew and Greek, slave and free, tax collector and zealot, male and female. This, in my view, is the key to the puzzle. The God who has been revealed here on Earth in Jesus is a God who overcomes differences and calls us together. In the presence of that God, all can sit together at the same table, in peace and fellowship, no matter who they are or where they are from.

It is not our place to decide whether ET can be a citizen of the Kingdom of God. It is our place to treat ET like the least one of

Christ's brothers and sisters—which is to say, to treat ET as we would treat Christ. And it is our place to live in the hope that ET will treat us likewise.

GUY: Remember George Coyne's insight that I quoted above, that, "in us, the universe has become self-aware"? By us, he meant humanity—with or without ETs. But really, even that's not quite true. Because in a sense there's no such thing as humanity, only individual people. It would be better said, in *me* the universe is self-aware. I carry with *me* the awareness of the universe.

As it happens, so do you. There are two of us, at least. By an interesting coincidence we are both residents of planet Earth and, thus, share a lot of DNA, but I can see that's rather irrelevant to the more interesting and important fact that we also both share the same attribute of self-awareness.

PAUL: In fact, there are a lot of us: billions and billions of us on planet Earth alone, billions more in the past. And, I suspect and hope, a lot more of us in the future—maybe not all on this planet, and maybe not all with the same DNA. But all of us—whatever planet or place, space or time we happen to inhabit—are the bearers of the purpose for which this universe exists. We are all at the center of the universe—from God's perspective. It was for us all that Christ died; it was for us all that the universe was born.

GUY: And the existence of all these other self-aware entities—you, or all the folks at the airport, or ET—raises an interesting challenge to each of us. (Especially high-tech introverts like me.) Are we willing to accept the presence of another self-aware entity in our universe? And even if we are aware of the presence of God, are we willing to accept that there are other self-aware entities, besides me and God, who are also aware of that same God?

It's the huge leap from having a nice, comfortable, one-on-one relationship, just me and God—what some people call being "spiritual"—to having to accept that other people are also partici-pating in that relationship. But that is what you have to face up to, when you belong to a religion. In fact, what we learn from the Gos-pels is that an essential part of our relationship with God is finding God in those others . . . be they our neighbors and family, or ET. (ET might be easier to deal with than some of my neighbors!)

Something occurs to me, standing here at the terminal at LAX. This airport cannot exist simply by itself. It only makes sense to build an airport if there are also other airports; there have to be places for the planes to fly to, places for the planes to come from. Airports are designed for communication.

But isn't the same true for civilizations, for religions, maybe even for intelligence itself? Intelligence only makes sense if there is someone else to share that intelligence with. We only grow and stretch ourselves when we're challenged to relate to others . . .

I'm an introvert, myself; I'm someone who would just as soon hide in my room with a book.

PAUL: But even a book assumes a dialogue between writer and reader. Without such a dialogue, no book would exist.

GUY: Certainly this particular book has been nothing but dialogue!

The ability to be self-aware, and the free will to act on that awareness, implies or maybe even demands the existence of an-other entity who is also self-aware, whom we can choose to love or to ignore. In that sense, ET would be no more alien, and no more scary, than the person standing next to me.

PAUL: Speaking as that person next to you, thanks. I think.

GUY: It's not a rigid proof but a pattern I am seeing. It suggests to me that it's not only on this planet that we are not alone. I expect that someday we'll learn that even in this cosmos, we are not alone.

DAY 7: Acknowledgments

PAUL: Before we take a day of rest, we should give our thanks to the people who helped make this book possible. We'll start with our agent, Gillian MacKenzie, and our editors, Gary Jansen and Amanda O'Connor. In addition, Bill and Kelley Higgins, Brother Bob Macke, SJ, and Jack Wisdom read over early drafts and gave us a lot of useful feedback.

GUY: The questions we've addressed over the past six "days" are not only frequently asked, they're frequently answered, as well. A lot of what we say here is material that we've presented in many places, in many ways, over many years. As I hope is obvious, the discussions in this book are significantly different from anything either of us has published before. But if you have happened to have read our previous work, that might explain why an illustrative analogy, formulation, or joke might sound familiar.

If you want to see a more formal version of my ideas on some of these topics, I can refer you to a number of articles I have written for the *New Catholic Encyclopedia* and its supplements, including entries about Astronomy, the Big Bang, Cosmology, Entropy, and the Universe. In 2012 I was invited to write an article, "The new physics and the old metaphysics: an essay for the use of Christian teachers," that was published in the journal *International Studies in Catholic Education*. After the Pluto decision at the IAU meeting in 2006 I wrote up a bunch of notes for my own memory; they formed the basis of our discussion in Day 2 but also went into an article I published in *The Physics Teacher* in 2007. Back in 2005 I wrote a little booklet for the Catholic Truth Society of London called *Intelligent Life in the Universe?* that included a lot of the ideas that come up again here in Day 6. And, of course, for the past ten years I've had a monthly column and other occasional articles in the British Catholic newsmagazine *The Tablet*, where a lot of my thoughts on these issues were first aired.

And while we're mentioning acknowledgments, we should also note that Scripture quotations used here are from the New Revised Standard Version of the Bible, ©1989 by the Division of Christian Education of the National Council of Churches of Christ in the USA. All rights reserved. Used by permission.

PAUL: Relevant to this book is my article "An Unblemished Success: Galileo's Sunspot Argument in the *Dialogue*," which appeared in the *Journal for the History of Astronomy* in 2000.

All the restaurants mentioned are real, except for the fictional restaurant of Day 5. And even there, the dishes we shared can all be found at several of our favorite places around Castel Gandolfo and Albano, near the observatory headquarters. The writing of this work was fueled in part by Italian cappuccino, often prepared by Guy in the coffee room of the Vatican Observatory.

GUY AND PAUL: All scholars stand on the shoulders of giants. We've tried to note in the text itself some of the main sources quoted or otherwise used in our discussions. In addition, we relied on other sources that didn't get cited directly. And, of course, you may want to follow up on any number of threads we left hanging loose. Below we provide a listing of some books that we think you might enjoy looking up yourself, if you want to follow our arguments further . . . or, at least, see where we stole our ideas!

DAY I: BIBLICAL GENESIS OR SCIENTIFIC BIG BANG?

When it comes to understanding cosmology, ancient and modern, many textbooks do a nice job of summarizing the material. One favorite for describing ancient science is A. C. Crombie's book *Augustine to Galileo: The History of Science A.D. 400–1650* (first published in 1952), which ignited Guy's passion for learning about the history of science. Also note Michael J. Crowe's *Theories of the World from Antiquity to the Copernican Revolution* (1990). If you want to read the original documents themselves, many important writings have been anthologized in Dennis Richard Danielson's *The Book of the Cosmos: Imagining the Universe from Heraclitus to Hawking* (2001). Another favorite for summarizing the medieval cosmology and putting it into its cultural context is *The Discarded Image: An Introduction to Medieval and Renaissance Literature* by C. S. Lewis (first published in 1964).

As for modern physics, we can recommend *The Quantum Enigma* by Bruce Rosenblum and Fred Kuttner (the second edition came out in 2011). Plenty of books describe Einstein's relativity in any number of levels of detail, but perhaps the easiest way to put that topic into perspective might be found not in science books per se but in biographies of the folks who developed those ideas, which also include good descriptions of the physics they were responsible for. In that regard, we personally like Abraham Pais's 1982 Einstein biography, *Subtle Is the Lord*, and John Farrell's 2005 biography of Father Lemaître, *The Day Without Yesterday*.

The history of the scientific revolution is discussed further in the Galileo chapter, but here we can mention Michael Buckley's *At the Origins of Modern Atheism*, published in 1990, which examines how the Enlightenment altered—not necessarily for the better—

our understanding of the role of God's action in the universe. An article-length version of Buckley's work (and lots of good articles by other scholars) can be found in *Physics, Philosophy, and Theology: A Common Question for Understanding*, edited by Robert J. Russell, William R. Stoeger, SJ, and George V. Coyne, SJ (1991).

For accessible scholarly reflection on the nature of Creation, from the perspectives of science, philosophy, and theology, see *Creation and the God of Abraham*, edited by David B. Burrell, Carlo Cogliati, Janet M. Soskice, and William R. Stoeger, SJ (2010).

Finally, we quote extensively from the 1996 address of Pope John Paul II to the Pontifical Academy of Sciences, "Truth Does Not Contradict Truth." This can be found online at http://www.newadvent.org/library/docs_jp02tc.htm.

Except for its mention in Pope John Paul II's address, we don't explicitly touch here on biological evolution—that's not our field—but certainly there are plenty of books that treat biology from a Catholic perspective. A good place to start is Kenneth R. Miller's *Finding Darwin's God* (1999).

DAY 2: WHAT HAPPENED TO POOR PLUTO?

The best description of Pluto and its "demotion" is found in Mike Brown's delightful and funny 2010 memoir, *How I Killed Pluto (And Why It Had It Coming)*. To learn more about meteorites and how they're collected in Antarctica, we have already mentioned Bill Cassidy's 2003 book, *Meteorites, Ice, and Antarctica*, and, of course, Guy's book from 2000, *Brother Astronomer*, which describes his role in one such expedition.

The point of departure for modern discussions of how science in general grows and changes over time—and in particular whether such change should be seen as cumulative or discontinuous—is Thomas Kuhn's classic *The Structure of Scientific Revolutions* (1970).

The idea of Pluto not as an ugly duckling but as a beautiful swan was suggested to Guy by an audience member at his Cranbrook talk. He doesn't know her name, but he would like to thank her here!

DAY 3: WHAT REALLY HAPPENED TO GALILEO?

There are so many Galileo books out there, it's hard to know where to stop. It's easy to know where to begin, however: *The Galileo Affair: A Documentary History,* a comprehensive collection of the documents in the case, assembled and translated into English by Maurice Finocchiaro (1989). Another source we relied on is David Marshall Miller's paper, "The Thirty Years' War and the Galileo Affair," published in the journal *History of Science*, volume 46 (2008).

Among other useful and interesting sources concerning Galileo are pretty much anything written by Stillman Drake; Richard Blackwell's *Galileo, Bellarmine, and the Bible* (1991); Annibale Fantoli's *Galileo: For Copernicanism and for the Church* (1994); and the various articles collected in *The Cambridge Companion to Galileo*, edited by Peter Machamer (1998). Our discussion was also influenced by Mario Biagioli's *Galileo, Courtier: The Practice of Science in the Culture of Absolutism* (1994).

A beautiful discussion of and guide to meridian lines and how they were used can be found in J. L. Heilbron's *The Sun in the Church* (2001). Owen Gingerich describes the history of Copernicus's famous text in *The Book Nobody Read* (2004).

DAY 4: WHAT WAS THE STAR OF BETHLEHEM?

A good astronomical explanation is provided in Michael Molnar's *Star of Bethlehem: The Legacy of the Magi* (2000). But there are plenty of other possible explanations; among the clearest is that presented is a short book by John Mosley, *The Christmas Star,* published by the Griffith Observatory (1987).

For a thorough discussion of the theology behind the infancy narratives, the classic book to read is *The Birth of the Messiah* by the noted biblical scholar Raymond Brown (the most recent edition dates from 1999).

A good and readable point of entry for putting religions and scientific accounts of the world into dialogue with each other is Olaf Pedersen's *The Two Books: Historical Notes on Some Interactions between Natural Science and Theology* (2007).

Day 5: What's Going to Happen When the World Ends?

The "setting" and background for this chapter were drawn from the fictional universe of Douglas Adams's comic work, *The Hitchhiker's Guide to the Galaxy* (1979) and its sequels.

In this chapter we touched briefly on only a few of the many ways people have suggested that the world could come to an end. An entertaining but scientifically legitimate collection of possibilities can be found in David Darling and Dirk Schulze-Makuch's *Megacatastrophes! Nine Strange Ways the World Could End* (2012).

Two books that are referenced in this chapter, which do a wonderful job of examining the role of religion in our technical society, are Jonathan Sacks's *The Great Partnership: Science, Religion, and the Search for Meaning* (2011) and Francis Spufford's *Unapologetic: Why, Despite Everything, Christianity Can Still Make Surprising Sense* (2012). While neither book touches directly on the issue of the end of the universe, both have good things to say about some of the not-so-peripheral topics that come up in our discussion on this day. Likewise, for further reflections on these themes we recommend John Polkinghorne's *The Faith of a Physicist* (1996) and John F. Haught's *Is Nature Enough? Meaning and Truth in the Age of Science* (2006).

We also make reference to Pope Benedict XVI's 2007 encyclical letter *Spe Salvi* (Saved in Hope). The text can be found in English online at http://www.vatican.va/holy_father/benedict_xvi/encyclicals/documents/hf_ben-xvi_enc_20071130_spe-salvi_en.html.

DAY 6: WOULD YOU BAPTIZE AN EXTRATERRESTRIAL?:

In the text, we cite two good sources on the history of attitudes toward extraterrestrials: Steven J. Dick's *Life on Other Worlds* (1998) and Michael J. Crowe's *The Extraterrestrial Life Debate* (2008). A recent source on the theological aspects, with a great bibliography, is Thomas O'Meara's *Vast Universe: Extraterrestrials and Christian Revelation* (2012).

We also make reference to Teilhard de Chardin's *The Human Phenomenon* (1955), which is well worth revisiting.